Risk Management: 10 Principles

This book is dedicated to my family who, over the years, have given 'risk management' a whole new meaning!

Risk Management: 10 Principles

Dr Jacqueline Jeynes PhD MBA BEd(Hons) BA

OXFORD AUCKLAND BOSTON JOHANNESBURG MELBOURNE NEW DELHI

Butterworth-Heinemann
Linacre House, Jordan Hill, Oxford OX2 8DP
225 Wildwood Avenue, Woburn, MA 01801-2041
A division of Reed Educational and Professional Publishing Ltd

A member of the Reed Elsevier plc group

First published 2002

British Library Cataloguing in Publication Data
A catalogue record for this book is available from the British Library

Library of Congress Cataloguing in Publication Data
A catalogue record for this book is available from the Library of Congress

ISBN 0 7506 5036 2

Composition by Genesis Typesetting, Laser Quay, Rochester, Kent
Printed and bound in Great Britain by Biddles Ltd, *www.biddles.co.uk*

Contents

List of figures

Glossary of main terms

- Hazard – something with the potential to cause harm or injury
- Risk – the likelihood that it will actually cause harm or injury
- Risk assessment – the process of identifying hazards and assessing the severity of harm and likelihood it will occur
- Risk factor – the range of factors that combine to represent the potential for harm, injury, damage or loss to occur
- Corporate governance – adherence to a set of principles to ensure proper controls are established and maintained within the organization
- Microfirm – up to ten employees
- Small firm – 11–50 employees
- Medium-size firm – 51–250 employees
- Large firm – over 250 employees
- HSC – Health and Safety Commission are the national body with the responsibility for considering health and safety issues and where the law may need to be amended to provide better or further protection for workers and others in the workplace
- HSE – the Health and Safety Executive answers to the HSC providing inspection and enforcement services
- COSHH – the control of substances that might be hazardous (that is with the potential to cause harm or injury) when used or stored, or disposed of. The substances can be liquids, gases, fumes, dusts and can be absorbed through direct contact with the skin, through breathing in, through swallowing and via other means such as through puncture wounds
- RIDDOR – Reporting of Injuries, Diseases and Dangerous Occurrences Regulations 1995
- Manual Handling – the action of handling large/heavy/awkwardly shaped/compact/uneven or sharp-edged objects (including people or animals). It relates to lifting, pulling, pushing or carrying these objects and the potential damage it can do to people if they handle things incorrectly. This can be lower back injuries, injuries to upper parts of the body and limbs and other injuries associated with dropping the object.

- CORGI – the central registration body for businesses and individual operatives working in the gas installation industry
- Control measures – an action/device/strategy intended to eliminate/alleviate/reduce the negative impact on the business or individual of a situation or event
- Direct losses – generally the more visible, more easily quantifiable losses that can be expected to occur and can be insured against to some degree
- Indirect or consequential losses – less easily quantified and less likely to be insurable.

Acknowledgements

I would like to thank all the people who have worked with me over the years, including colleagues at the Federation of Small Businesses, and who have been good enough to listen to me developing my thoughts on the '10 Ps'. In particular, I would like to thank Stephen Fulwell for his ideas about the spider diagrams to highlight where priority risks are, and Tony Briscoe from IBEC in Ireland for his suggestions about the role of planning and performance in the equation.

Part 1

Chapter 1

Introduction

1.1 Aim of the book

The last five years of the twentieth century witnessed significant changes in the way firms operate and in the fundamental structure of business units as globalization became more prominent. British industry has changed from primarily manufacturing based to predominantly service provision and, after the mergers and takeovers of the 1970s, came the trend for down-sizing to much smaller business units in the 1980s and 1990s. Total number of businesses has grown to around four million, the vast majority being sole traders or partners without employees, accompanied by the rapid growth in the use of telecommunications, the internet, part-time and temporary employment contracts and the use of home-working.

Membership of the European Union has brought with it a stream of legislation and, more recently, a desire to bring all member states into closer alignment on employment and worker protection, social issues, taxation and other fiscal measures. This has been closely followed by many directives which seem to be blurring the edges between different disciplines when transposed into national legislation. Despite greater emphasis on recognizing the needs of small firms, there are considerable pressures, both internal and external, that require firms to be able to demonstrate to others that they are managing the business satisfactorily.

While Figure 1.1 identifies some of these pressures, when considered alongside the changing and uncertain face of current competitive climate, we can see why risk management is often sidelined in smaller organizations.

The ten elements of operation that represent the main risk areas to the success of a business are considered to be:

1 *Premises* – where the firm is located, type of premises available for use, amenities, distribution routes, access for customers
2 *Product* – industry sector, features of product or service offered, life cycle and fashion trends, materials used in production, green issues, quality

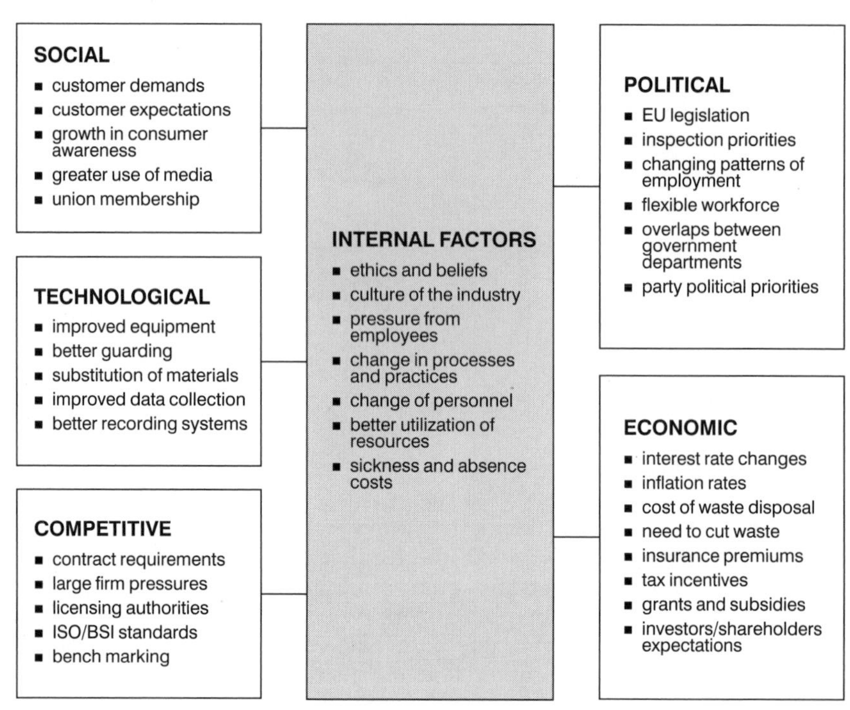

Figure 1.1 Internal and external pressures on business

3. *Purchasing* – access to supplies, storage and warehouse facilities, stock control, payment terms, cost
4. *People* – the workers in the organization, skills, training needs, motivation and commitment, incentive packages available, employment contracts
5. *Procedures* – production procedures, record keeping and reporting systems, monitoring and review, use of standards, emergency procedures
6. *Protection* – personal protection of workers and others, property and vehicle security, insurance cover, information systems, data security
7. *Processes* – production processes, waste and scrap disposal, skills, technology and new materials
8. *Performance* – targets set, monitoring, measurement tools, consistency, validity of data
9. *Planning* – access to relevant data, management skills, external factors and levels of control, short- and long-term planning, investment options
10. *Policy* – range of policies that support the strategic plans of the firm.

Each element represents its own type of risks that interact with, and impact on, the others sometimes positively and sometimes negatively. No-one can eliminate all the risks in all the areas – it is a risky business setting up, operating and developing a successful operation. With careful management they can, of course, be reduced or controlled sufficiently to alleviate or spread the risk, hopefully in such a way that retains the excitement and challenge of running a successful business while protecting all stakeholders from potential harm.

An evaluation of all the business operations requires honesty and motivation in order to produce a comprehensive, detailed analysis of potential risks within all ten areas listed above, referred to as the 10 Ps. A daunting task, but a necessary one in order to gain a true appreciation of how all the elements fit together, rather than the 'sticking plaster' approach to dealing with risks piecemeal as they materialize.

The 10 Ps approach outlined here considers each of these ten areas of business management for the risk factors and controls in place, providing prompts and tools for assessing the risks they present in order to prioritize subsequent risk reduction activities required. These risk factors include issues such as:

- employment – related to employing workers, need for skills, management structures, shortages, employment protection
- legislation – including discrimination law, health, safety and fire protection, environmental protection, permits, procedures, record keeping
- security – safeguarding people, premises, data and copyright protection, theft, violence to staff and others
- competition – pricing strategies, location, bench marking and standards, public perceptions, penalties
- finance – investment, insurance and litigation, returns and profit, long- and short-term planning.

Risks are allocated a rating against such factors as:

- extent of potential harm or damage
- likelihood it will occur
- possible disruption to business activities or growth
- short- or long-term effects
- internal strengths and weaknesses
- ability to recover
- likely impact on owners/shareholders/public image
- litigation.

Clearly a wholly theoretical approach is of limited value when, in reality, businesses do not operate in such a nice neat way! However, it is vital that all risks to the business should be considered strategically at the most senior level, not just financial risks, and an approach that can be used consistently throughout is a valuable tool for management. Of particular significance now is the emphasis on visible corporate governance demonstrated through adherence to the Combined Code of the Committee on Corporate Governance[1] from December 2000. Transparency is the key word, whether related to government activities or the business world, so ability to produce evidence of actions taken to safeguard the interests of stakeholders is critical.

The structure and size of the organization will impact on the depth and breadth of risk management activities required, as will the industry sector. It is also important to note that judgement will be needed at an individual level and this book is not intended to be an over-arching requirement on every firm irrespective of its relevance. It is intended to provide comprehensive guidance to those who have responsibility for ensuring adequate corporate governance at board level, for managers responsible for establishing and monitoring procedures to support the strategic objectives of the business, plus others who provide advice and guidance to the business community.

1.2 Business structures

As noted earlier there are vast differences in the way businesses are organized that will impact on how the proposed approach is used. It is useful to consider the underlying assumptions and beliefs of the author that support the development of this approach to risk management. These are that:

- despite the stated goals of reduced burdens on business and 'better regulation', new legislation introduced in recent years has had a much more fundamental impact on the way business operates than any of the regulations that have disappeared
- larger organizations are generally (not always of course) better placed to accommodate such legislative changes than smaller firms, so the burdens as cost per employee are often disproportionately applied – in some cases as much as 10 per cent more
- growth in the number of small firms will continue in the near future
- consumers expect more, particularly on environmental protection and being seen to operate an 'ethical' business at global level
- workers expect more in the way of protection from health and safety risks, a greater say in major business decisions and more direct consultation
- there is greater emphasis on accountability and higher expectations of results from those in senior management positions
- external factors play an ever-increasing role in how the company is perceived by customers, shareholders and other stakeholders.

Despite evidence to suggest otherwise, there still exists the view that small firms are just scaled down versions of large organizations. With around 94 per cent of private-sector firms in the UK employing fewer than 10 people, it is vital to acknowledge the differences in organization that such a microfirm requires. However, as business structures continue to change, the definition of 'small firm' based on the number of employees becomes less helpful, though still an important indicator. It would seem that a combination of number of employees/ industry sector/turnover/incorporation status might be more helpful when considering potential risks and management priorities to control them.

This is not a book about management theory in general, but clearly there are some points worth noting relative to business structure and the range or type of risks that might have the greatest impact. The following summary presents the main features in this context against the most common forms of business entity, from sole trader to plc.

1.2.1 Sole trader/self-employed individual

Usually unincorporated with a fluid, flexible management approach, which suggests that potential risks can be spotted and dealt with more quickly. Although entrepreneurs are often considered to be risk-takers, this is more likely to be 'calculated risk' taking, using less formal methods for analysing and evaluating risks. The biggest problems are likely to be related to:

- lack of knowledge and awareness about legislative requirements
- restricted access to finance and poorer financial management skills
- concentration on production rather than administrative or management issues
- higher ratios of insurance premium
- restricted access to some markets and reliance on a limited customer base
- less formal methods of monitoring and control
- poorer quality premises and plant.

The owner is the critical person and the central pivot of the organization's culture if and when workers are employed.

1.2.2 Partnership

Shared responsibilities, skills and financial investment between two or more people, plus potential for employing other staff. Apart from the obvious risk of one or more partners running off with the assets, risks are likely to be similar to those of the self-employed individual. Potential difficulties relate to authority, control, decision-making and monitoring controls. Significant risks may also be evident in small, high-tech enterprises, especially new firms with high-cost borrowing and little business experience. Partners may operate in an even or uneven collaborative way and the industry sector may present greater risks associated with process or licensing requirements.

1.2.3 Small private limited companies

Up to around 50 employees or a small business unit as part of a larger organization, these tend to be hierarchical in structure, with culture and beliefs established by the original owner. Often in traditional industries or sectors, although a growing number are high-tech or service sector firms.

Risks often associated with:

- poor communication channels
- need to change management approach as the firm grows and the early flexible style needs to become more structured or formalized
- inappropriate premises or facilities to support growth of the firm
- product life cycle and product development
- lack of training or facilities to develop up-to-date skills
- inability, or unwillingness, to change to meet challenges of modern competitive environment.

1.2.4 Medium-size limited companies

Larger concerns, more visible to a wider band of customers, therefore there are greater risks associated with consumer choice, environmental protection policies and public image. Risks also associated with insufficient communication and feedback channels resulting in too little or too much information to maintain an effective risk management programme throughout the firm. While shallower hierarchical structures are developing, larger organizations inevitably have to devolve power and authority to smaller business units, adding to the potential base of data feedback but also potentially to confusion.

Risks may be diverse and impact on various divisions, increasing the need to have systems in place to manage risks effectively. Such organizations are often slower to recognize and react to financial or competitive risks and there is significant potential for operational risks to be ignored or given insufficient weighting when considering overall risk management strategy.

1.2.5 plcs and large organizations

The larger the organization, the greater the need for formal risk management strategies that flow from the top down. In addition to the risks already outlined above, internal risks are associated with lack of coordination of specialist department interests and information channels, making it more difficult to take a holistic view. There has to be greater reliance on feedback of relevant information at senior level, with many opportunities along the way for dilution or amendment of data.

Recent legislative changes mean greater transparency is needed, with wider dissemination of business information, wider worker participation in decision-making, plus eco-friendly pressures from consumers. Public image often becomes more significant as the organization grows and risks associated with public or financial market perceptions increase as ability directly to control or reduce such risks decreases. At this level, the issue of risk-sharing and the balance of potential insured losses against uninsured losses needs careful management. Commitment and motivation may present additional risks for firms in merger or takeover situations.

Though not exhaustive, the above summaries highlight some of the potential risk factors that firms must consider irrespective of sizes, and the need for a systematic approach that is relevant, comprehensive and cross-functional, while acknowledging the unique spread of pressures facing individual firms.

1.3 10 Ps of risk management

The risk management approach identified by the author in *Practical Health and Safety Management for Small Businesses*[2] was developed specifically to help those working in or with small and microfirms. It considers the health, safety, fire and other legislative risks to the business. While primarily concerned with physical or visible risks, thus making it easier for non-specialists to establish a workable management approach in the absence of any other, it provides a useful base from which to develop a more holistic system for analysing and evaluating business risks in general. By adding further elements of planning and performance measurement it becomes a much more comprehensive management tool that can be tailor-made to suit the individual firm.

Figure 1.2 shows how the different elements impact on each other, and although these 10 principles cover the main elements comprehensively, it is hardly a nice easy number to remember! They have, therefore, been broken down into four distinct groups of:

1 Physical properties – *premises/product/purchasing* supplies
2 People elements – *people/procedures* they follow/*protection*
3 Actions or processes – *processes/performance* against targets
4 Management issues – *policy* and strategy/*planning* and organizing.

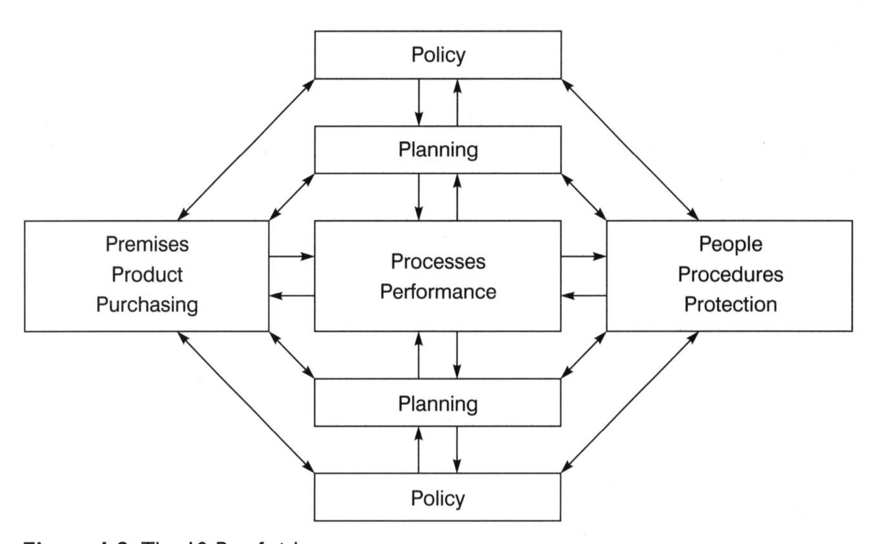

Figure 1.2 The 10 Ps of risk management

These all overlap or interact with each other constantly, so cannot be separated out too far. However, they do provide a structure from which to identify and evaluate risks to the business, and to initiate and monitor controls to reduce these risks.

While many text books focus on the policy of the firm as a starting point, in reality businesses tend to start from the concrete and move on to strategic issues in a more untidy, organic way. Wherever the reader chooses to start within the 10 Ps structure, they will inevitably move backwards and forwards to the policy and planning elements, especially smaller firms. At the larger end of the scale, senior management will rely on input from others at business unit or division level, who may also choose a different starting point.

The next section, Part two 'Identifying risk factors', considers possible risk factors against each of the ten headings, allocates some form of risk rating against each and decides on priorities for action. Part three will develop this further as various options for control measures are explored.

Part 2

Chapter 2

Identifying risk factors

2.1 Risk assessment

As the current health and safety (H&S) legislation in Europe depends on a risk assessment approach to managing and controlling hazards, there is a great deal of information and guidance available on what this involves. While this book is not exclusively concerned with health and safety risks, there is a legal requirement to carry out such assessments so it is prudent to start from this point.

The principles of risk assessment are quite straightforward, based on the following activities:

1 Identify hazardous conditions/properties/processes that could potentially cause harm, injury or damage
2 Consider what this harm, injury or damage might be; who could be affected; and how serious the result of exposure might be
3 Evaluate the likelihood that such harm, injury or damage will occur, taking into account any control measures that exist
4 Make judgements about adequacy of controls in place, identify gaps in adequate provision and prioritize actions needed to correct the situation
5 Monitor and re-evaluate after appropriate time scales and when circumstances/materials/processes etc. change.

Despite the fairly simple logic of this approach, there has been much hype and confusion generated about what *risk assessment* actually is, with the result that the potential value and practical application of the process has become lost under a mountain of paperwork. This does not mean that records of the process are unnecessary. Clearly such activities should be recorded in some form to confirm they have been carried out adequately and to ensure that:

1 the scope or extent of activities to be assessed is clearly identified beforehand
2 the full range of potential hazards or risk factors has been considered

3 people in the organization know what these are, what controls are in place and how to use them
4 adequate monitoring and review can take place
5 and other parties can see that risks are being managed appropriately.

Indeed, significant findings of health and safety risk assessments should be recorded by law, certainly when five or more people are employed. It is particularly important that risk assessments are relevant to the business itself, are carried out by suitably experienced, competent people, are of sufficient depth to ensure people and property are protected and that they reflect what actually happens in the firm rather than what management thinks should be happening. This does, therefore, require input from several sources, both users of the systems as well as managers, technical experts or health and safety specialists.

In the author's view, the least successful approach is the use of external specialists to 'do risk assessments' for the firm with little or no input from internal staff. There may well be valid reasons for using external consultants for some elements of the risk assessment, but this should be identified through earlier assessment activities within the firm. Even the specialist expertise required for assessments such as noise levels, fibre/air concentrations, or asbestos materials will need to involve workers directly.

For some readers, this will be familiar ground and H&S risk assessment will have been a regular occurrence as it has been a legal requirement for at least 8 years. Other readers may be familiar with evaluating broader risks to the business but not specifically health and safety. In addition, changes to the fire precautions regulations mean that many more firms are now brought into this requirement than previously, as emphasis is now on a risk assessment rather than a fire certificate prescriptive model and all firms employing more than one person have to carry out a fire risk assessment.

The intention in this section is to establish a risk assessment approach that can be applied to all the elements of the 10 Ps, bringing together the fire and H&S risks with other issues and concerns that need to be controlled and managed in the organization. The starting point in Chapter 2 is to identify hazards within the firm using a practical checklist approach, then to evaluate the hazards or risk factors in Chapter 3 and to allocate risk ratings in Chapter 4.

2.2 Identifying hazards

In all industries there are hazards or risk factors that are so glaringly obvious to those working with them that they are hardly worthy of note. Unfortunately, the frequent reference to 'common sense' belies the previous knowledge, skills and experience of individuals that help them to recognize such hazards and makes it less likely that newcomers to the industry are made aware of them.

Example: A young man, Simon Jones, was killed within hours of starting work at Euromin (based at Shoreham Docks) in 1998, after being sent there by an employment agency. He was given no training in how to carry out the tasks set, which were particularly hazardous given his position in the cargo hold in relation to the crane grab being used and was disadvantaged still further by the total disregard for any safe working procedures.

In larger organizations, different sections or divisions may well have a base of knowledge of hazards but no central source of reference. In addition, results may be patchy, depending on the way data has been collected and the range of people involved in its collection. It is important that a structure is agreed and applied throughout the firm to make it both consistent and easier to collate, of course, but also to ensure all the relevant risk factors have been identified. Records should also include details about where and when the assessment is carried out and by whom.

A valuable starting point is a site plan (Figures 2.1 and 2.2), irrespective of size of firm, as it gives visual reminders about areas of activity that are sometimes forgotten, such as external waste storage areas and rarely-used storerooms. It is useful for highlighting movement of people and goods through the firm and potential areas for conflicting priorities of use. From an insurance point of view it also demonstrates that a comprehensive analysis has been carried out and that the planning and policy elements are based on relevant information.

In addition, a checklist for movement of goods through the business such as the one included here (Figure 2.3) provides a useful structure to work to and includes elements of delivery of materials as well as onward travel. It can work just as well with a service type of business as a production one, in that movement through the business may be of people, animals, or objects. The list of stages can be amended of course to reflect the work of the particular business unit and should reflect what actually happens on a day-to-day basis. It also enables people involved in these different stages to confirm the type and range of activities carried out fairly quickly.

The areas to consider include:

- Car parks, delivery areas, pedestrian access
- Reception and waiting areas
- Storage of supplies and materials
- Movement of goods from stores to first process stage
- Office and administration areas
- Public display areas
- Work-in-progress stages
- Maintenance and repair
- Activities carried out off-site, including movement between sites
- Storage of waste/scrap materials
- Packaging and storage before distributed
- Distribution.

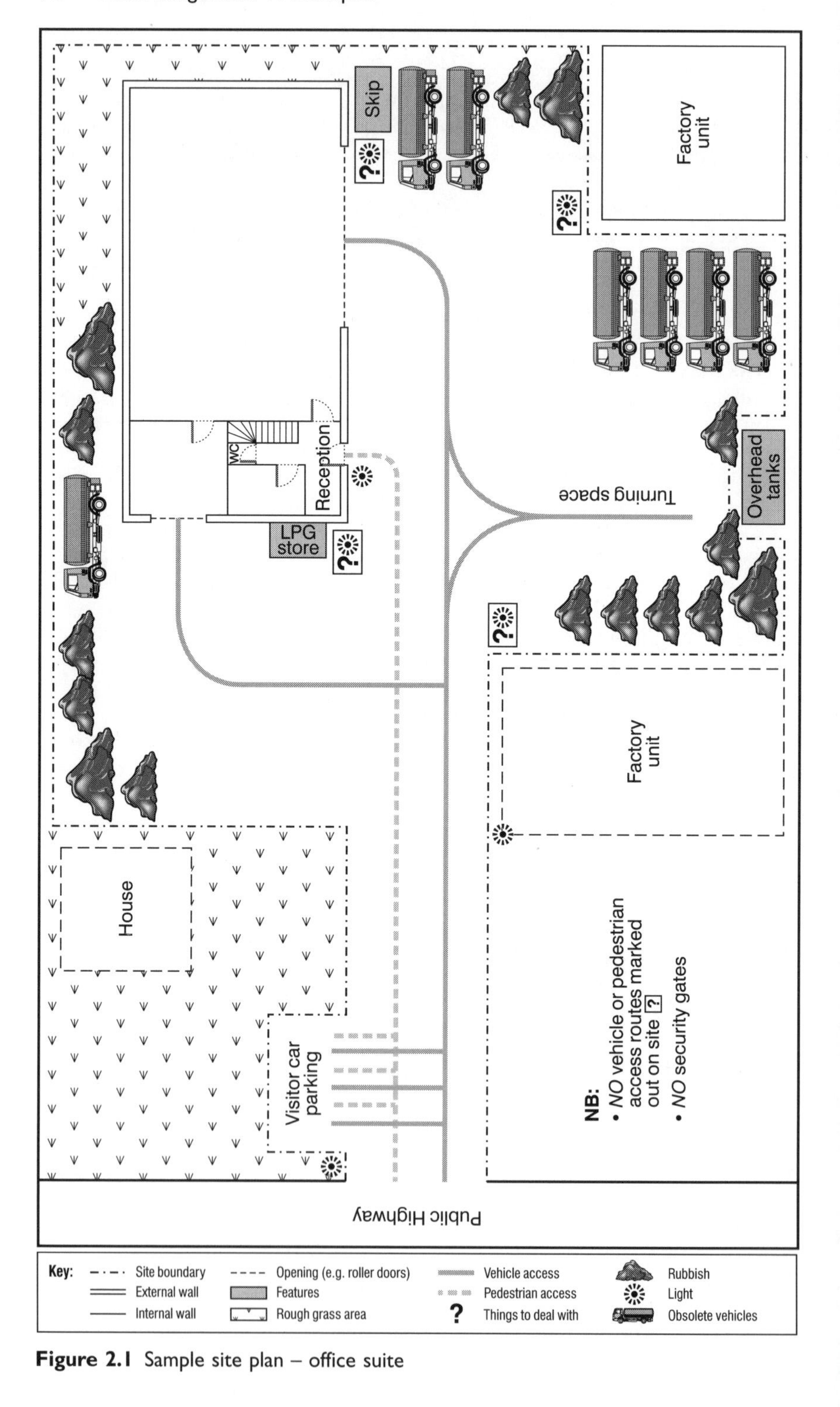

Figure 2.1 Sample site plan – office suite

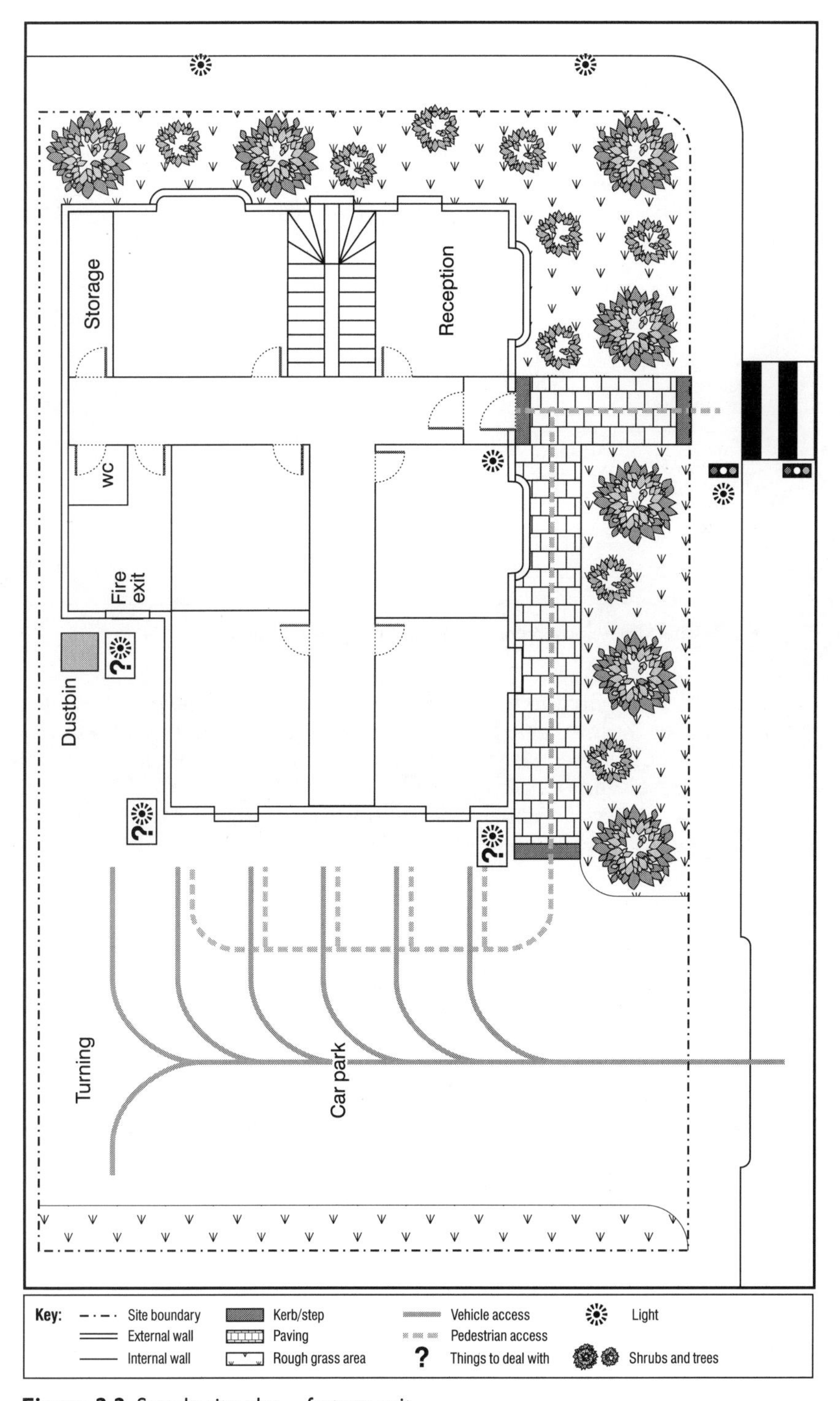

Figure 2.2 Sample site plan – factory unit

Site location: ....
Date of assessment: ....
Name(s) of assessor(s): ....

Stage of progress	Main activities	Equipment and processes used	People involved	H&S issues	Security issues
Stage 1: Arrival					
Stage 2: Process					
Stage 3: Process					
Stage 4: Completion or finishing					
Stage 5: Onward movement					

Figure 2.3 Checklist – movement of goods through the business

It may also be valuable to consider what equipment or processes are used at each of these stages, especially for health and safety and security risk factors considered later and the number of people involved (including those who only occasionally need to be present).

The following section considers each of the 10 principles against the risk factors identified in Chapter 1, within the four groups of:

(a) physical properties
(b) people elements
(c) actions or processes
(d) management issues.

2.3 Risk factors:

(a) Physical properties – premises/product/purchasing

2.3.1 Premises

This is often a significant risk factor for smaller firms, as they frequently have limited access to suitable premises either at start-up stage or when expanding production. At the other end of the scale, larger concerns with a variety of sites have additional risk factors to consider and must fully optimize the facilities available at each. A wide range of risk factors include:

- suitability of premises to type of process involved, particularly in traditional manufacturing industries
- size of premises and facilities available
- financial concerns related to ownership or tenancy status, repairs, expansion etc.
- location and means by which the product or service reaches the customer
- health, safety, fire and environmental risks to workers and others.

Employment risk factors

- Location – skill shortage areas and access to suitably qualified or experienced staff, particularly in areas of traditional industries now redundant
- Relocation – need to relocate critical staff members; costs involved and employment packages required for whole family relocation

Example: Loss of major industries such as mining, heavy manufacturing or banking, represent significant risk factors to firms in the area. The base of highly skilled labour from these industries may require substantial retraining for work in different sectors.

- Capacity of premises for workers and customers
- Capacity for production processes and people to operate them
- Provision of facilities for staff, customers.

Legislative risk factors

- Equal opportunities and other discrimination laws relative to potential workforce base
- Safety – structure of buildings including narrow or uneven stairs and steps; layout of buildings for moving goods or materials; movement of vehicles on-site; electrical supplies; adequate lighting and potential for slips and trips; working areas considered to be 'confined spaces'

Example: Older premises are not easily converted to reduce potential safety hazards of narrow or uneven stairs and steps. Modern purpose-built commercial premises may still present potential problems with changed production processes or greater use of computerized systems.

- Health – heating and lighting provision; ventilation and clean air; smoking of workers or customers; washing and toilet facilities; extreme heat/cold areas of working

Example: Increased pressure to provide smoke-free working environments may not present problems where spare capacity is available for providing specified smoking areas. On the other hand, some premises are not suited for such segregation, especially leisure areas where customers expect to be able to smoke.

- Health – the requirement to identify and record all areas of the site where asbestos materials may be present; damaged or exposed sections of asbestos materials; potentially hidden sources of asbestos, particularly older premises
- Fire – structure of buildings including open-plan areas and stair-wells; access routes; escape routes/facilities available in emergencies; structural materials; location relative to other high-risk premises; shared premises and multiple occupancy; need for fire certificate

Example: Shared premises may present additional hazards associated with safety or security of common areas, such as lobbies, stairs or escape routes. It is particularly important to consider processes of others in shared premises, proximity and additional fire hazards introduced by their activities.

- Environment – disposal and storage areas for hazardous materials; proximity to residential areas regarding noise emissions; ventilation and exhaust systems used; removal/disposal of asbestos materials used in structure

Example: Use of asbestos materials in the structure of buildings needs to be identified, potential risk to workers and others assessed and removal, where necessary, to be carried out under controlled conditions.

- Permits or licences to operate in given premises; licence requirements to safeguard workers and public; planning permissions.

Security risk factors

- Visitors to the site and control of access to certain areas; fencing and gates

Example: Access to the site may or may not present additional risks, but unauthorized entry to shop-floor areas by delivery staff, for instance, may introduce further hazards.

- Unauthorized visitors – potential for injury; arson; property damage; vandalism

> Example: Potential for arson is more difficult to identify in some situations than others. Piles of broken pallets against buildings near to perimeter fences are ideal targets for arsonists, as are waste skips full of flammable materials. Less obvious targets may be closed or 'secure' storage areas that require more effort and planning on the part of the arsonist.

- Theft of materials; goods produced; money and other valuables; workers/customers' valuables; safe storage areas
- Access to electronic data; personal records; confidential information
- Access to harmful substances/agents; storage of hazardous substances and materials
- Violence to staff, visitors, animals on site.

> Example: Consider areas where staff may be particularly vulnerable at certain times, such as opening up in the early morning or locking up at night, when working alone, or when moving money and valuables around the site.

Competitive risk factors

- Declining or growing commercial locations – impact of nearby derelict or empty premises on customer perception
- Changes to local by-laws – local authority changes to parking and loading facilities; changed one-way road systems; diverted road and pedestrian traffic

> Example: Environment and transport policies are being produced by all local authorities in the UK and these are likely to include restricted access to town centres at certain times of day; specified times for loading or unloading vehicles; charges or tolls. Introduction of traffic calming measures involves road closures and potential impact on passing trade.

- Customer expectations of premises, image, facilities
- High-street or out-of-town location.

Financial risk factors

- Cost of purchase, duties, legal costs
- Local authority and utilities charges and potential changes to these

> Example: Recent serious flooding (during 2000) in many parts of the UK not usually affected has already resulted in massive increases in insurance premiums, limited insurance cover, or no cover at all for future flood damage. Uninsured losses can be up to ten times more than insured losses. In addition, local authorities are expected to increase council tax and business rates significantly in order to recover costs of clearing up and improving flood defences.

- Terms of leasing agreements; timespan for agreement renewal
- Life span of plant; mark down and write-off values
- Maintenance requirements; shared premises and responsibilities
- Charges for car park spaces
- Insurance premium rates dependent on growth or decline of local environment
- Additional security precautions required due to location, such as CCTV or metal shutters.

2.3.2 Product or service

> There are several risk factors associated with the product or service itself that then feed into questions about purchasing, production, waste management etc. These include:
>
> - stages of the life cycle of the product and recent trends
> - the firm's competitive position now and potential in the future
> - 'green' environmental issues that affect development
> - life-style trends and demographic changes
> - a range of internal and external pressures on the business.

Employment risk factors

- Skills of workforce; relevance to current requirements and trends
- Age structure of workforce and demographic trends

> Example: General ageing of the work population means fewer young people in some industry sectors, such as nursing or care sector where the most experienced and skilled groups of workers are expected to retire during the next decade. More workers are now caring for older relatives, as life expectancy increases, as well as children or young adults who are living in the parental home for longer.

- Willingness and ability of workforce to up-grade their skills
- Need for flexible staffing due to variation in demand; contract conditions
- Staff ratios required

> Example: Also a legislative and competitive issue, the ratio of staff to client/customer may present problems, for example in childcare services. Working time regulations require an 'uninterrupted' break of 20 minutes in a 6-hour shift which may require additional staff cover for limited periods during the day.

- Skill levels and experience of sales force.

Legislative risk factors

- Employment contract conditions and employment law protection
- Working patterns and hours (working time legislation); holiday and break entitlements
- Product safety – potential for injury if not used properly; labelling and packaging; sharp edges and/or moving parts; weight/size/shape of product; child-safety requirements

> Example: As UK society follows the US lead and becomes more litigious, so potential hazards associated with the product or service need to be carefully considered. Providers of information, guidance and advice are just as susceptible to such factors as producers of consumer products.

- Consumer protection – use of non-hazardous materials; description of product or service and customer perception (see also litigation later)
- Worker safety – use of substances and materials; sharp edges and moving parts of product; personal safety when working with the public or animals and potential for violence
- Health – biological contaminants during production; fumes when in production or during storage; allergenic or carcinogenic properties

> Example: Agriculture and horticulture, scientific testing facilities, food production, and medical services are all sectors that need to identify the potential risks to health from exposure to biological or chemical substances. This is important whether dealing with animals or humans.

- Fire – product acts as a source of ignition/fuel/oxygen when in production, use or storage; flame retardant properties; testing requirements; potential for explosion; flammable qualities

- Environment – methods of distribution if hazardous; disposal of product; obsolescence; substitution of materials for production or packaging
- Licence requirements for provision of service/production/movement and distribution/disposal of waste and scrap materials; specified qualifications for workers.

> Example: Recent regulations have been introduced for transporting hazardous goods or materials, with specific systems and qualifications required at all stages of the journey.

Security risk factors

- Copyright and registered designs; patents
- Research and development security; use of technology in production or materials used for product/service
- Use of temporary workers; fraud; industrial espionage potential
- Unauthorized access to data; restricted access to parts of production process
- Theft; fire; damage; storage of high-value components/work-in-progress/or finished goods
- Protection of money and other valuables; amount of money on premises at any time
- Security of vehicles and personnel in transit between sites.

> Example: Often overlooked, the security of vehicles is a risk factor whether they are stored on-site or used by staff to travel to and from work. Delivery of materials, components and finished goods should be considered, as well as transport of equipment or plant to carry out the job off-site (such as construction).

Competitive risk factors

- Life-cycle stage of product or service; consumer trends and cyclical demand

> Example: Apart from obvious short-term fashion trends, there are longer-term life-cycle risk factors for most products. For example, the licensed trade and restaurant industries have been particularly badly hit in recent years, but coffee bars have seen a massive rise in popularity in just three years.

- Demographic changes relative to product or service
- Pricing range; sales methods used

> Example: Use of the internet as a method for reaching customers is growing daily, particularly in retail and as a means to 'cut out' the agent or broker in transactions, such as house purchase, holidays, or financial services. Even small firms traditionally trading in small local areas can use this sales method, or can view it as a further risk for their business.

- Target quality levels; quality control and assurance methods
- Use of standards and bench marking in industry/firm
- Changes to legislation regarding use of materials/products, both UK and EU
- Public perception of industry and firm; bad publicity for other businesses in industry.

> Example: The rail industry is a good example of the potential risk to an organization from bad publicity within an industry having an impact on individual firms. It may not be easy to forecast what the triggers might be for such a lack of confidence, but the potential risk could be considered on a 'worst scenario' basis if necessary.

Financial risk factors

- Profit ratios; costs of production rising/steady/falling
- Range of suppliers available/used; payment and delivery systems agreed
- Storage and warehouse facilities and costs; call-off facilities; use of Just in Time for materials or components; stock control; penalties
- Delivery methods and costs; cost of fuel

> Example: The year 2000 witnessed considerable financial damage to many firms as the cost of fuel escalated, shortages in supply occurred and transport of goods was disrupted. This affected many industry sectors and firms with little or no contingency planning in place were at a distinct disadvantage.

- Collection of payments; debt/credit control systems; electronic transfer of payments; payroll systems
- Cash flow; banking facilities; lending rates; expected return on investments
- VAT and other tax regimes for firm's products or service
- Licence costs; worker training and qualification costs

Example: The gas industry has been particularly hard hit financially by changes to the licence and registration scheme operated by CORGI, the introduction of new qualification structures for operatives and increased call on resources to fund the relevant training.

- Cost of investments; repair and replacement of equipment/machinery/buildings.

2.3.3 Purchasing

This is a significant element in the management of risks that is often isolated from consideration of the other elements. There are broad issues such as:

- the use of recognized standards in the business
- the firm's policy on quality
- government policy on standards, environment, protection of workers etc.

plus more specific issues for the firm such as:

- cost and payment conditions
- types of materials, availability, delivery
- production processes and techniques
- technology and renewing or replacing equipment and machinery
- 'green' issues and public perceptions of the firm.

Employment risk factors

- Skills and experience of purchasing team; ability and resources to consider best options for purchase
- Existence of effective communication between all parts of the firm to ensure appropriate ordering programme
- Access to information about needs of different departments; awareness of needs of any management system standard in place
- Need for very close liaison with planning departments and dialogue on planning issues.

Example: The risks associated with mismanagement of the purchasing or procurement function may be significant. The larger the organization, the more potential for problems related to communication channels and inappropriate ordering schedules. This may be exacerbated by central purchasing policy that does not take full account of individual business unit requirements.

Legislation risks

- Competition laws and the need to operate a fair procurement system

> Example: A high proportion of firms do not review purchasing arrangements to consider the optimum option, but stay with the same suppliers for many years. While such loyalty sometimes results in preferred supply conditions, the risk factors associated with this approach relate to requirements for changing supplies to match changing processes or methods of production; the ability of the supplier to continue operating as planned; and potential for restricting access to other suppliers unfairly.

- Payment systems and issue of late-payment
- Safety – adequate storage facilities; safe loading/unloading areas; use of vehicles on site; storage and handling of hazardous substances and materials; retention of Hazard Data Sheets
- Health – manual handling of loads; use of vehicles for handling; noise levels; light, heat and ventilation in storage areas

> Example: In industries such as agriculture and horticulture, it is still difficult to receive supplies in size of container and quantity that reduces risks of manual handling injuries.

- Environment – transport of materials and products; training and qualifications for handling or transporting some materials; arranging waste disposal services
- Systems for monitoring location and state of materials; records of disposal
- Replacing plant and machinery with safer/more environmentally friendly/more efficient/quieter versions.

Security risk factors

- Checking supplies in to confirm quantity/quality; checking materials out to relevant people
- Theft; fire; damage; vandalism; arson
- Safe storage of hazardous substances; authorized access
- Pilfering and theft by staff/others
- Movement of goods/materials between sites; transport safety and security; use of refrigerated or other specialist vehicles and containers

> Example: While there has been a significant shift towards automated sales transactions in recent years, there are still many cash-based businesses such as leisure or retail. Risks may, therefore, be considerable for some firms, including potential for injury to staff as well as loss of money (often uninsurable).

- Movement and storage of money and valuables
- Perimeter fencing; delivery vehicles and staff.

Competitive risk factors

- Pricing strategy; access to appropriate materials and suppliers
- Use of environmentally-friendly materials; public perception

Example: Substitution of hazardous substances in printing, for instance, may represent competitive risk factors if sufficiently robust or effective substitutes are still not available. For example, the use of water-based inks for printing on flexible plastic products results in print that does not always 'fix' satisfactorily – hence blue hands from carrying the printed carrier bag home!

- Suppliers or customers introducing different management or control systems; need to up-date procedures and reporting systems
- Introduction of new legislative requirements for industry.

Financial risk factors

- Rising costs of supplies
- Rising fuel and transport costs; vehicle and other licence costs
- Public pressure to substitute some materials/components of production
- Research and development costs to use substitute materials effectively; staff retraining
- Costs of insurance for secure storage
- Opportunity cost of holding large quantities or high-value stock.

(b) People elements – people/procedures/protection

2.3.4 People

It is important to consider workers at all levels in the firm, especially those with non-traditional forms of work contract and temporary workers. There are broader considerations for some firms, as risks to visitors to the site and the wider public in the vicinity may need to be identified. Other issues include:

- how workers are organized, for example as groups or teams
- cultural issues including the 'culture' within individual workplaces
- whether there exists (or should exist) union recognition for workers
- skills and competence of current workers and how closely these fit future needs
- training and supervision of workers
- legislative requirements aimed at reducing risks to workers.

Employment risk factors

- Range of skills and experience of current workforce and match with activities
- Skills and expertise required for future work and activities; gap between these two

> Example: Training is a crucial element of maintaining the skill base and the type of skills training required by the firm may not easily be accessed locally. Risks are also associated with the format or structure of available training, or the ability to analyse exactly what is needed.

- Level of retraining needed; cost and time involved; access to new staff
- Need to find relevant training provision; internal or external provision; training and support available internally
- Amount, type, quality of supervision and management of workers; organization of work groups and communication links between them
- Culture of organization; review and feedback mechanisms available
- Recognition of union membership
- Opportunities for promotion and development of staff; barriers to advancement; facility for staff to move between departments
- Use of temporary staff; motivation and commitment of staff.

> Example: High staff turnover and use of a large proportion of workers on temporary contract may represent an additional risk. This may become more apparent when linked to the need for providing adequate health and safety training for staff.

Legislative risk factors

- Equal opportunities, racial discrimination, disability discrimination requirements; reflection of balance in local community

> Example: Some industries find it difficult to recruit relevant skilled and qualified staff that reflects the make-up of the local community, so leaving themselves open to charges of non-compliance with relevant anti-discrimination legislation.

- Provision of state-defined benefits, maternity and parental leave etc.
- Work patterns; working time legislation and access to time off; rest and washroom facilities

> Example: In nurseries and other childcare premises, staff have traditionally taken lunch breaks with the children under their care, rather than separately as an uninterrupted break. Risk factors related to these legislative requirements relate to staff deployment, additional cover for breaks and financial implications of this in the fee structure for customers.

- Minimum wage and pay structures
- Safety – adequate training and supervision; provision of relevant information; use of adequate protective gear/equipment; safe equipment, machinery, materials; regular maintenance procedures; when driving safely, on behalf of firm; lone working
- Health – health screening and surveillance required; lighting, heating, ventilation; use of VDUs; sight and hearing testing may be necessary; exposure to hazardous substances; potential air contamination and biological hazard exposure; extremes of temperature
- Exposure to violence and/or stress; working from home; working in other people's premises
- Staff ratios for some tasks.

Security risk factors

- Violence to staff from customers/other staff/unauthorized visitors

> Example: Violence or abuse of staff by members of the public has grown considerably over recent years and must be acknowledged as a risk factor by any organization that deals directly with the public.

- Use and movement of money and other valuables on and off site
- Theft and damage by workers or others; fraud and access to confidential information
- Data protection; use of information technology systems
- Unauthorized access to parts of site
- Security of company vehicles on and off site; private use of vehicles.

Competitive risk factors

- 'Poaching' trained staff by other firms; recruitment strategies
- Commitment and motivation of staff
- Access to sufficient suitably-skilled staff
- Demographic trends and ageing population; lack of young people entering industry

Example: The gas industry is an example of an ageing, skilled workforce with significant reduction in the numbers of young people entering the sector. Risks are then associated with loss of skills as older workers retire and a shrinking skill base of workers to take their place.

- Payment and financial incentive schemes.

Financial risk factors

- Cost of wages and other payments; ratio of wage bill to sales; profit sharing schemes

Example: Rising costs of employment represent a risk factor for firms, particularly those with high staff turnover. Changes to 'industrial injuries' support structures will present further risks as employers will need to establish rehabilitation systems for injured or disabled workers.

- Minimum wage; rising national insurance contributions (NIC) for employers; taxation levels
- Recruitment costs; turnover rates for staff in different positions in firm
- Cost of provision of state benefits through payroll; cash flow
- Direct and indirect costs of skills training; costs of H&S and other relevant training
- Insurance premiums; costs of injury and ill health to workers; provision of occupational health protection schemes; potential provision and cost of rehabilitation services
- Sickness absence; cover for other periods of absence; maternity benefit provision.

2.3.5 Procedures

This element relates to others in the 10 Ps quite closely, particularly the product, process and people. The sort of questions you might consider should include:

- how appropriate are they for current production processes?
- will they be appropriate for future production?
- how will the introduction of new technologies impact on existing procedures?
- are they actually implemented as they should be and are they monitored effectively?
- how is their effectiveness measured and evaluated?
- do they serve to reduce risks or pose additional ones?

Employment risk factors

- Lack of relevant skills in present workforce
- Inexperience of workers; high proportion of newcomers to industry, for instance from redundant industries in region; significant proportion of 'vulnerable' groups of workers, especially young people

> Example: Sectors with high staff turnover rates, such as the leisure and catering industries, may also have a high proportion of young workers at any one time. Potentially, such workers are more vulnerable due to lack of experience and skills, but also because they have not had time to learn a job fully before moving on. However, mature workers with high-level skills from old traditional industries may also be more vulnerable when having to change career direction.

- Complacency in use of long-established procedures; development of 'short cuts' to inappropriate procedures

> Example: All industries can experience complacency among staff where procedures have been in place and remained unchanged for a long time. In production areas this may lead to 'short cuts' to speed up the task, for instance by over-riding safety mechanisms. This complacency may also relate to inaccurate record-keeping and reporting procedures that then represents a risk to the validity of data used to support management decisions.

- Access to relevant skills training; need for different skills in future and resistance to training
- Need for fewer low-skilled workers; increased use of technology to reduce number of people employed per business unit; redeployment of staff and resources.

Legislative risk factors

- Safety – changes in law regarding use of materials/substances; greater specification of safe methods of working in some industries
- Need to analyse procedures for risks to individuals, including safe working practices and permits to work; monitor, supervise and control use of procedures on day-to-day basis

> Example: Safety hazards must be identified for individuals where necessary, whether for short-term or long-term working. For example, in the short term there may be additional risk factors for pregnant or nursing women, or someone recovering from an accident or illness. Long term, the additional hazards present for individual left-handed workers operating machinery designed for right-handed people – with stop buttons placed wrongly – are often overlooked.

- Health – working time; exposure to stressors; workloads; noise generation and protection for workers
- Ensure changes to procedures do not introduce additional hazards for workers or others
- Record-keeping procedures; number of people involved in collecting and collating data; relevance and accuracy of data collected; internal and external reporting procedures, e.g. RIDDOR
- Emergency procedures in the event of fire, explosion, spillage or leakage of substances
- Environment – transporting goods or substances; disposal of hazardous waste; licences for movement and disposal; records; qualifications of relevant staff; notification systems.

Security risk factors

- Emergency procedures in event of theft, violence, damage, arson
- Safeguarding sensitive or confidential data
- Security of IT data and systems; illegal use of internet access by staff or others

> Example: As IT systems become more complex and sophisticated in their operation, so too do the means to access them illegally. This may therefore represent a risk factor, whether it relates to medical or sensitive personal information about clients/patients/workers, or financial data of individuals.

- Movement of money and other valuables; site and vehicle security
- Systems to monitor movements of staff, particularly lone or mobile workers.

Competitive risk factors

- Use of out-dated, slower methods of production
- Need to replace plant and equipment with more efficient models; ability to meet customer demand for new features or facilities of product or service
- Customer service expectations; replacement/refund/compensation procedures

- Minimum/optimum/maximum production schedules and ordering systems
- Use of third party certification schemes for quality/environment/occupational health and safety management systems (such as ISO 9000/ ISO 14000/ BS 8800 standards)

> Example: For firms without third party certification systems in place, these may represent a risk factor if clients use them as a means of restricting access to goods or services. As a user of such schemes, there are several risk factors associated with their use, including time and human resources to establish and maintain the system/cost, often becoming a net cost to the firm/difficulty in integrating 2 or 3 different standards for quality, environment and OH&S and of reducing impact of conflicting or overlapping requirements of these systems.

- Use of industry standards or codes of practice; recognition by customers.

Financial risk factors

- Cost of implementing and maintaining third party certification schemes
- Cost of replacing out-dated or inefficient plant and equipment
- Short- versus long-term investment programmes; potential returns on investments and time scale

> Example: The 'boom and bust' cycles experienced in the UK have left some negative equity situations, substantial losses on share values in some sectors and potential problems for those responsible for making financial borrowing or investment decisions.

- Cost of training and retraining staff; cost of recruitment
- Time and resources needed to monitor use of procedures and maintain adequate records where needed
- Insurance and potential litigation; balance of insured and uninsured losses.

> Example: Insurance protection generally only covers around 20 per cent of the real cost of incidents and, indeed, many people are already under-insured. Operational changes may not have been notified to insurance provider for some time and values for plant may be much higher than original premiums allowed for.

2.3.6 Protection

This is much broader than just protection of people from health and safety risks and includes identifying risks associated with the protection of:

- people
- premises
- materials
- intellectual rights
- data and security
- the environment

plus other concerns such as insurance and the law.

Employment risk factors

- Generation and protection of ideas by workforce; opportunities for workers to put forward ideas and take part in consultation with management
- Protection of jobs; use of flexible working patterns; maintaining motivation and commitment of workers
- Use of different employment contracts; conflict of definition of 'worker', 'self-employed' person etc. and expectations of worker protection

Example: As worker protection measures increase, so too do risk factors associated with balancing these measures against the need for a flexible workforce and the needs of different groups of employees on different employment contracts.

- Pension provision; other benefits and incentives.

Example: From October 2001, all firms employing more than five people will need to provide workers with access to some form of pension scheme as the norm, rather than as an optional extra. The government's Stakeholder Pension provision should be available from April 2001 to support this.

Legislative risk factors

- Employment law requirements re dismissal, redundancy, time off, contract of employment, discipline and grievance procedures etc.

> Example: The base of case study material to demonstrate the impact of the Human Rights Bill is not yet sufficient to evaluate its impact as a risk factor. It is likely, however, to impact more on labour-intensive industries or those that are highly-regulated at present.

- Statutory Sick Pay; Statutory Maternity Pay and benefits; minimum wage; correct application of collection/payment of government schemes
- Safety – provision of adequate training, information and supervision; provision of safe working areas with safe tools and equipment; safety gear and personal protective equipment (PPE); regular maintenance and repair programmes; protection of vulnerable groups of workers

> Example: Some sectors, such as construction or agriculture, will need to reduce the residual, inherent hazards of the task as much as possible before relying too heavily on personal protective equipment for individuals.

- Health – health surveillance monitoring; recording results; confidentiality; use of results for management and/or human resources decisions; rehabilitation programmes in-house

> Example: Access to occupational health services is a growing requirement in the UK, as it is in many European countries and is likely to become a mandatory requirement for most workers in the future.

- Fire – greater emphasis on protection of surrounding areas in the event of fire; impact on people/ buildings/land/animals

> Example: The recent review of Fire Safety Regulations has seen a shift of emphasis to a risk assessment approach, with specific references to the need to consider the potential impact on, and protection of, the local environs if a fire occurs.

- Environment – protection of internal and external environment; emissions; economic use of power sources; use of renewable sources where possible.

Security risk factors

- Protection of people and premises; use of barriers to keep people away from hazards
- Unauthorized access to plant and machinery

- Unauthorized access to records and protection of individual's human rights
- Provision of appropriate training and support for staff to deal with potentially violent situations.

> Example: Industries where face-to-face contact with the public is involved, crime rates and instances of threatened or actual violence to staff have increased in recent years, posing additional threats to personal security and safety.

Competitive risk factors

- Cost of providing wide range of protective measures; need to keep up-to-date with changing conditions
- Resources required to maintain systems and thus taken away from primary production
- Public perception of image of company; ethical approach; environmental protection

> Example: Potential loss of credibility or positive image with customers may be a risk factor for firms as a result of actions by others in the industry. This may relate to methods of selling, such as Time Share properties or double-glazed window units, or to the product/service itself, such as incorrect advice to clients on pensions during the 1980s and 1990s.

- Industry image, positive or negative; impact of bad publicity attached to other firms in locality or industry
- Uneven distribution of legal compliance among main industry players; inappropriate use of schemes to restrict access to suppliers
- Growing base of employment protection and other government-led measures represent burdens on firm (may be more relevant to size of firm)
- Need for transparency of decisions balanced by need for commercial confidentiality.

Financial risk factors

- Cost of employment protection measures; difficulty of replacing key staff when absent (for whatever reason) and additional costs involved

> Example: Larger organizations may be able to accommodate the absence of key staff members for some time, although the costs might be substantial. This is more difficult in smaller organizations, those that are traditionally female-dominated (with potential maternity absence), or in larger concerns with high levels of absence in certain divisions.

- Increasing redundancy cost burden on employer and, potentially, lack of adequate provision for future
- Balance between costs of different employment contract terms
- Interest rates on planned borrowing for investment.

(c) Actions or processes – process/performance

2.3.7 Process

Risks associated with the process itself can vary enormously, of course, depending on the type of business being considered. However, the fundamental questions will be related to:

- the techniques used and inherent risks associated with them
- controls in place to reduce risks
- potential impact of technological developments, both positive and negative
- changes in legislation and their impact on choice of techniques
- government initiatives to support and encourage firms to consider using new technologies
- skill levels of available staff, both in-house and more widely available in the geographic area

Employment risk factors

- Lack of skills; lack of motivation or commitment
- Direct contact with the public often leading to high levels of stress; potential or actual violence to workers
- High turnover of staff due to unsatisfactory work conditions or requirements

Example: Whether the firm employs many or few workers, they are a crucial element of providing the product or service to the customer. Retaining a committed workforce is, therefore, vital, so organization of the process should support this. The volume of work and the way workers are organized are significant factors, as are access to relevant information and input to decisions that affect them directly.

- Increased workloads and insufficient staff numbers; too many people and not enough work to gainfully occupy them; fluctuations in customer demand/seasonal work etc.
- Little opportunity to take control or make decisions about process used
- Inadequate information, training, supervision
- Repetitive, low-skill work
- Team or group working; piece work; targets for production.

Legislative risk factors

- Disability discrimination – possible adjustments to make work accessible to people with some form of disability
- Job descriptions and skill requirements matching tasks and processes actually in place
- Use of physical parameters for suitability for work
- Safety – safe equipment and machinery to ensure efficient working; appropriate use of physical guards; elimination or substitution for most hazardous parts of process; dealing with residual hazards of the process and ensuring safety of workers or others

> Example: Greater reliance on computerized systems has generated different hazards for workers, with health issues becoming more prominent. Examples include constant use of telecommunications equipment at call centres, or long periods of time at computer screens.

- Health – use, handling, storage of hazardous materials and substances; air quality and emissions at different stages of the process; regular checks and system to investigate concerns of workforce; manual handling; use of VDUs; use of biological agents
- Environment – emissions and exhaust systems; noise levels; spillage or leakage of substances.

> Example: Traceability of substances into the local environment is a priority for various enforcement bodies, so must be considered a potential risk factor of production processes. Consideration should include deliberate disposal, such as waste products from food production, as well as potential for accidental spillage or leakage of chemicals or biological agents.

Security risk factors

- Copyright, patents, registered designs; industrial espionage (intentional or unwitting)

> Example: Copying high-profile branded goods is a multimillion pound business world-wide, especially fashion items and sports gear. Protecting brands and designs may therefore represent a considerable risk factor.

- Damage or theft of materials, goods during production or distribution
- Safety of goods in transit; storage facilities
- Unauthorized access to commercially sensitive or extremely hazardous areas
- Taking part in bench marking or third party initiatives with potential competitors in same industry.

Competitive risk factors

- Out-dated processes and lagging behind competitors
- Reliance on suppliers of components
- Transport and delivery costs; infrastructure weaknesses in region/country-wide

> Example: Global warming, regional and national transport policies and increased flooding have all had a negative impact on infrastructure, representing a risk factor for distribution and delivery sectors.

- Increasing use of IT and internet methods of communication
- Product life-cycle stages; access to research and development facilities
- Public perception and changes in moral stance (such as the use of fur for fashion clothing)
- Differences between requirements of local authorities (LA) when operating across different LA areas; international differences in regulatory requirements.

> Example: Where firms operate across national or local authority borders, different regulatory requirements may present competitive and financial risks. Devolution, regional changes and the introduction of Regional Development Agencies may add to these risk factors.

Financial risk factors

- Insurance costs and potential for payouts in future if current acceptable processes found to be hazardous

> Example: Processes and substances in use for many years have later been found to be hazardous to workers and others, such as wood dust, and subsequently led to litigation. While no-one can foresee whether this will be an issue, there may be some indications evident now. For instance, vapours given off during certain stages of flexible plastic processing may be seen as just an irritant now, but may later be classified as harmful.

- Litigation for damage or harm to workers, visitors to site, local environs
- Investment time scales, particularly for rapidly changing technology
- Public and financial market perceptions of company's use of processes
- Access to affordable finance for investment.

2.3.8 Performance

As a risk factor, this relates to the criteria and performance measures chosen by the firm. Who are the stakeholders who actually want to know about performance and what are these different groups actually looking for? Clearly, these questions will then impact on the type of measures chosen for a specific element of business performance and how effectively risks are actually being managed.

Performance can be viewed at individual worker/department/company level and may just be related to the individual firm or be part of a bench marking exercise. Again, questions of health and safety, accidents and injuries, insurance claims, quality and environmental standards will all be part of the evaluation of risk management.

Employment risk factors

- Are performance measures appropriate for the range/type of work undertaken; are they relevant, meaningful and actually represent measurable targets?
- People are involved in setting performance targets for themselves or the team; everyone knows what they are

Example: Performance measures must reflect what people do to complete the task and therefore must have input from workers themselves. Where people have negative experiences of trying to reach unrealistic performance measures, they tend to be more sceptical of others that are introduced. Good communication skills are vital.

- People are trained in the use of performance measures; feedback is provided when and where it should be
- Relationship with discipline and grievance procedures; are they used as a positive rather than a negative management tool?

Example: The scope and range of disciplinary issues stated in employee handbooks sometimes become so great as to be a charter for dismissing as many people as possible in the shortest space of time! On the assumption that employees are necessary for the firm to operate, performance measures should be decided in order to optimize their performance in order to help the firm achieve its overall objectives, rather than as a route to inevitable failure.

Legislative risk factors

- Employment protection laws, especially dismissal
- Safety – reliance on historical data about accidents/sickness absence/near-misses; need for 'no-blame' culture to ensure reporting is accurate; difficult to set targets that can be linked back to positive safety performance

> Example: Historical data on accident rates per annum or sickness absence, for instance, may be patchy or incomplete. Any data collected need to be analysed closely in order to provide indicators for future targets. Investigating accidents is an area often neglected in firms, though a legal requirement, especially those with a blame culture that results in punishment for anyone recording accidental damage to people or property.

- Health – similar issues to safety performance; long delay between symptoms appearing and actual event that caused them; difficult to measure accurately or find root cause; need for accurate, probably specialist, measurement data
- Fire – fire prevention and fire fighting performance can only be practised in simulated situations (for example fire drills)

> Example: Emergency drills for dealing with a fire, restricting its spread and evacuating people can only be tested for validity in the event of a real fire. Many people do not take fire drills seriously, often do not know what the procedure is and may not have experienced one while at work – for instance, few pubs or restaurants ever hold a practice evacuation of the premises even though their customers may be very vulnerable in the event of fire.

- Environment – may be long delay between cause and effect; can be immediate and catastrophic; positive performance sometimes difficult to measure; scale of penalties and financial levies.

> Example: Government levies introduced in recent years will have an impact on high energy-users and the 'polluter pays' is the strongest message coming through in new regulations for environmental protection.

Security risk factors

- Confidentiality of personnel assessment reports and feedback
- Breach of confidentiality on company-wide or divisional performance inside and outside organization

> Example: Firms with stock market listings can suffer serious damage from lack of adequate security on information. Even smaller organizations not listed can suffer damage from leaks of sensitive information about future plans or developments to the local media.

- Leaks of information to markets or media.

Competitive risk factors

- Poor results in bench marking activities
- Bad publicity from breaches of third party certification standards; withdrawal of certification

> Example: While third party certification schemes can offer valuable 'badging' opportunities for firms, inability to maintain the standards set or actual withdrawal of the certification is a risk factor to be considered. This might relate to quality MSS such as ISO 9000 series for example, or registration to operate such as the CORGI registration for gas operatives.

- Poor market reactions and subsequent loss of confidence by stakeholders
- Public perception of company as an employer if poor safety or health performance.

Financial risk factors

- Cost of maintaining performance measuring systems; time and other resource costs
- Cost of accidents and injuries if poor safety performance; increased insurance premiums

> Example: The cost of accidents is underestimated in firms that do not have direct experience of incidents at the workplace. While direct costs may be fairly obvious, indirect costs related to loss of production time, impact on witnesses, management time to investigate accidents and impact of bad publicity locally are often overlooked.

- Withdrawal of financial backing; loss of trading value on world stock markets; higher cost of lending
- Cost of penalties and fines, particularly for environmental performance
- Loss of sales if poor performance; reduced profits; reduction in planned investment; reduced production capacity.

(d) Management issues – planning and policy

2.3.9 Planning

This includes planning at the strategic management level and the practical operations level, with all the other elements feeding information back to this stage and priorities for action being decided. While all the other Ps feed into this stage, it is vital that they are considered in such a way that they are all equally important in the decision-making process. For example, much of the work over recent years to develop tools to help businesses manage health and safety risks is due to this element being given a much lower priority than financial issues. Sometimes this may be justified, but recent shifts in the emphasis of regulations covering the workplace mean that all risks must be considered.

Questions at this stage include:

- what is the purpose of the planning activity – and who is interested in the results?
- who will be involved in the planning process, either internal or external to the firm?
- how do all the other Ps feed into and impact on this stage?
- how will priorities for action be decided?

Employment risk factors

- Insufficient input from all levels within the firm; exclusion of some groups of workers, especially those working on shifts/off site/travelling around from site to site
- Little understanding of the planning process and little commitment to the findings
- Balance between needs of different sites or business units

Example: As firms expand direct contact with each operational element of the business reduces and delegation of authority and responsibility is necessary. Mergers and takeovers may also bring together business units with different beliefs and cultures, making it difficult to take the necessary holistic view of employment issues.

- Lack of knowledge or understanding of issues at operational level by senior management
- Lack of knowledge and awareness of current legislative requirements

> Example: Legislation changes very quickly, so it is vital for businesses to keep up-to-date with changes. This involves time and cost and takes away effort from the primary function of providing a product or service. Access to information in a relevant format is important for all firms.

- Ability to predict future skills needs of firm.

Legislative risk factors

- Deciding priorities for action that acknowledge the legal requirements on business; time scales involved
- Changes to licensing requirements for processes/procedures/qualifications
- Different legal requirements at regional LA level and between countries
- Conflict between different enforcement bodies on acceptable actions required to comply with the law

> Example: As regulations become more complex and far-reaching, there is sometimes more scope for conflicting requirements from different enforcement bodies on the same issues, such as health and safety, fire and planning requirements in food preparation sectors.

- Changes to law that require alteration to premises; problems associated with age/type of premises and ownership status

> Example: Changes such as those within the Disability Discrimination Act (Premises) and the Fire Precautions (Workplace) Regulations may require alterations to premises. The age, type and structure of the buildings will be a major factor, of course, but so too will be the ownership status or responsibilities of the occupier. Multiple occupancy premises may have particular difficulties in this area.

- Amount of record-keeping and monitoring activities required by law.

Security risk factors

- Access to relevant data to assist with planning

> Example: Collection and analysis of data from different sources or divisions within the organization may be time consuming and fraught with problems, although this may be reduced to some extent through the use of electronic means for transferring data. Security of these data inside and outside the firm may pose additional risks, particularly if speculative ideas are being formulated in the early stages.

- Inappropriate use of data, either internally or externally.

Competitive risk factors

- Correctly identifying consumer trends and product life-cycle stage
- Awareness of competitors' plans for the future
- Developments in the industry; technological developments; access to relevant research

> Example: Information and communications technology is advancing so quickly that it is very difficult to plan realistically for future requirements. In larger organizations with many different business divisions and interests, the balance between the need for central control of these diverse elements and the individual needs of each division is a crucial risk factor.

- Ability to access relevant skills and expertise to assist in planning process
- Relevance and sufficiency of available data; timeliness; volume of data to be collated and fed into planning process.

Financial risk factors

- Resource implications to support proposed plans for future
- Cash flow and access to necessary funds; opportunity cost of investments; prioritizing

> Example: As so many firms go bankrupt with healthy order books for the future but a critical shortage of funds in the short term, the cashflow is crucial. Late payment by businesses has been blamed in the past, and the government's Late Payment bill has tried to alleviate this to some extent. However, the payment system of the firm and its clients/suppliers is still a financial risk factor of all planning activities to consider.

- Rapid depreciation of some assets
- Government actions in future that impact on long-term planning programme

- Base rate/interest rates for borrowing/tax rates and incentives/VAT and other duties
- Exchange rates/strengths and weaknesses of world currencies; strength of sterling

Example: Firms trading outside the UK will be aware of the financial risks associated with exchange rates and the strength of the pound against other currencies recently. The question of Britain's stance on the European currency (Euro) and proposed harmonization on taxation and VAT, for instance, will continue to be a risk factor during the next few years.

- Consumer spending trends at home and abroad; inflation; regional upturns or collapses
- Uncontrollable events external to firm – e.g. fuel price rises or shortages.

2.3.10 Policy

This may have been placed first on your own list on the basis that theoretically this should be the starting point. However, in practice this is not necessarily where people begin, certainly not in smaller firms where practical considerations dictate many of the subsequent policy decisions. It is, of course, a critical element in developing strategies that will enable the policy aims to be met.

A more realistic scenario may be one where various other elements in the 10 Ps list feed into policy discussions and decisions, thereby making it a more dynamic element. Various policies may be developed in relation to:

- health and safety
- accident investigation, reporting and rehabilitation
- environment and waste management
- employment and equal opportunities
- purchasing and financial control
- competition.

Clearly such policies must be developed in such a way that they coexist easily with others and to have the greatest impact must build on feedback from the other elements identified. In addition, the question will always be 'how do these policies affect the identified risks?'

Employment policies

- Involving and consulting with employees/workers on issues that affect their work
- Providing sufficient resources, information and training to enable them to do their job effectively

- Ensure equal opportunities for employment and advancement in the company, irrespective of gender, ethnic origin, religious beliefs, age, disability
- Safeguard workers from bullying or harassment in the workplace
- Setting targets for individual, team, division achievement that are achievable, measurable and realistic
- Provide adequate supervision and management at all levels of firm, with regular feedback on work
- Comply with all relevant employment protection requirements
- Provide relevant discipline and grievance procedures.

Legislative policies

- Providing a safe and healthy work environment for people
- Produce a product or service that does not jeopardize the safety and health of others or the environment
- Ensure plant and machinery are maintained properly, good housekeeping standards are set and maintained and adequate facilities are available for those on site
- Establish processes and procedures for work in all areas of the business that take into account the health and safety of workers; ensuring these procedures are followed correctly
- Provide adequate protection against the risks of injury, harm or damage resulting from work activities or fire; ensure the protection of vulnerable groups of workers, including lone workers
- Provide a 'clean air' environment based on a smoking policy; no alcohol or illegal drugs allowed on working premises; policy on drink/drugs and drivers
- In event of fire, safety of people will take precedence over damage to property
- Policy on working time regulations.

Security policies

- Individual right to security of personal records and restricted access to confidential information; permission from individual to pass information to third party
- No commercially-sensitive information to be passed to others outside the company
- Protection of designs
- Identifying and reducing opportunities for workers and others to be placed in personal danger of attack (such as locking-up/paying cash to bank)
- Policy on theft from company, clients or other workers.

Competitive policies

- Purchasing less hazardous and more environmentally-friendly materials where possible
- Using local suppliers within *x* miles radius of business unit

- Appropriate use of third party certification schemes to ensure quality of supplies, without introducing unfair and restrictive requirements
- Ensure any industry standards and codes of practice are adhered to
- Ensure appropriate record-keeping systems are in place and maintained correctly
- Establish and maintain effective financial controls to safeguard stakeholder interests, taking into account relevant legislative requirements
- Identify and control the range of risks that can impact on the business.

Financial policies

- Ensure proper financial systems are in place to meet the firm's financial commitments to workers, suppliers, shareholders and others
- Provide sufficient resources to ensure all legal requirements are met satisfactorily
- Policy on late payment of bills
- Establish systems to reduce potential for fraud within the organization
- Ensure adequate internal controls to meet all obligations
- Provide sufficient insurance cover for insurable costs; take action to spread impact of uninsurable costs.

Chapter 3

Evaluating the hazards

Clearly, the previous section is an extremely comprehensive list of the potential hazards or risk factors that any type or size of firm may find. While there may be elements of each individual item that applies to any organization, the assumption is that at a business unit level, the scope of factors will be much smaller and therefore more manageable.

However, it must also be noted that this is only the starting point for evaluating the risks to the organization and that it is important to spend sufficient time at this stage to ensure full coverage of the potential risk factors. It is then much easier to justify removing or combining some of these elements at future stages of the process in order to control and manage the risks effectively.

3.1 What results are likely from exposure to these factors?

So far, we have identified 'hazardous conditions, properties and processes that could potentially cause harm, injury or damage' as noted at the beginning of Chapter 2 and to some extent considered what form of harm, injury or damage could result. It is important to look at the range of hazards or risk factors in detail to get a clearer picture of what the results of exposure to the hazard could be, how serious the resulting harm, injury or damage might be and who is most likely to be affected by it. In some cases, this will be specific to individuals at the shop-floor or office level, in other cases it will impact directly at business unit or divisional level and in others the resulting damage could affect the whole organization or locality in which it operates.

Potential harm, injury or damage may therefore be related to:

- Employment factors – such as inability to recruit suitably qualified and experienced staff, leaving the company vulnerable to poor workmanship and quality problems; a workforce made up of many people soon to reach retirement age but with few young workers entering the industry; inadequate premises that are poorly located, so customers are discouraged from visiting the site; low morale or commitment of workers, high staff turnover and potential for needs of vulnerable groups to be ignored.

- Legislative factors – such as increased costs and time involved in compliance with employment legislation, especially where internal knowledge of the law is limited; personal injury to workers or customers from unsafe actions; harm to individuals and the local environment from the incorrect handling of products, leading to increased calls on insurance cover; changed priorities of local authorities in enforcement actions or planning decisions, calling for increased investment by company; insufficient safeguards against the start or spread of fire, causing potential damage or injury sometimes over a considerable area.
- Security factors – this is growing in importance for many firms as the potential for harm can be considerable, ranging from leaking sensitive information that can fundamentally damage future investment plans of the firm, to loss of personal data of workers or customers; theft, burglary and violence to staff are increasing in some regions, leading to increased cost of physical guarding and alarm systems, as well as potential claims for injury or loss.
- Competitive factors – such as declining industry sectors, the need to compete more globally than previously and investment costs of research and development, all leading to potential loss of sales, profits and market share; consumer pressure for more environmentally friendly products or processes may require significant changes to operations; use of third party certification schemes may be positive or negative for the organization, depending on the industry sector and perhaps the size of firm.
- Financial factors – increased cost of operating the business, recruiting and retaining relevant skilled staff, have to be balanced by other costs; late payment is critical to cash flow and can significantly increase costs of borrowing short-term; reliance on small customer base leaves the firm vulnerable to the loss of just one or two customers; overall increased emphasis on protection and compliance with a wide range of legislative requirements takes funds away from the direct purpose of the organization, that is to provide a specific product or service for the customer; external perceptions about the firm can have major financial impact when stock market values are volatile.

As we have used the same five risk factors against each of the 10 Ps, the process will inevitably have led to repetition or overlap of factors. Having carried out this vital stage at such depth, the risk factors can now be grouped together before considering the potential impact of the hazards. Tables 3.1–3.4 summarize the risk factors identified earlier, breaking them down into more manageable chunks. The breakdown of headings in Column 1 should reflect the type of factors found in the individual firm, so this is just an example of how it might be done. It is still useful at this stage to keep to the four groups of factors representing physical properties, people elements, processes and management issues. The examples identify typical factors that relate to a manufacturing plant.

Column 5 can be used to pull out the factors that occur across the group, such as skill base in the local area, or facilities available and may help to emphasize significant factors that appear regularly and need close attention.

Table 3.1 Risk factors identified: group (a) premises/product/purchasing

Risk factors	*Premises*	*Product*	*Purchasing*	*Common features*
Employment: location/skills/ demographics	Access Size/capacity Facilities Car parking Image	Age of workforce (10% to retire in 10 years) Mid life-cycle stage of product Distribution routes Local skill base	Distribution routes Few local suppliers Poor communication link with production	Location and distribution channels
Legislation: employment/ health & safety/ environment/ other	Lack adequate facilities Age premises/stairs/corridors Unsafe structure Heating/lighting Noise: close to residential area Restrictive planning requirements	Fire risk re use of materials Need for safer substitute materials Product safety features Disposal when obsolete Amount/type of packaging used	Need to source safer materials Poor facilities for drivers Insufficient storage Disposal of waste	Inadequate storage facilities Need for new sources safer supplies
Security:	Safety in local area Safe storage of vehicles Warehouse facilities poor	Copyright protection Storage and warehousing Access to data Pilfering small parts	Perimeter fence damaged and stores not fully secured Checking-out system not working	Local security measures inadequate Internal security systems
Competition: industry/ consumer/internal	Unreliable power supply locally Poor local amenities Growing sector/no purpose-built premises Image: nearby derelict shops	Decline in traditional industries locally Short-term life span so constant up-date needed New R&D facilities needed	Turnaround time for deliveries increasing Packaging Limited access to eco-friendly substitute materials	Packaging and disposal Image/R&D
Finance: internal/external	Increasing maintenance costs Cost relocation of key staff Cost insurance and security Incentives for staff	Increasing cost materials Cost of scrap High cost packaging re price Increased transport costs	Increased cost of transport Late payment by customers Our late payment restricting supply from main supplier	Cost of materials Cost of insurance Fuel costs

Table 3.2 Risk factors identified: group (b) people/procedures/protection

Risk factors	*People*	*Procedures*	*Protection*	*Common features*
Employment: location/skills/ demographics	Low levels of skills locally Need for training High level temporary work contracts used	Worker inexperience Recording procedures not fully working	Flexible working and need for staff cover for absence Unauthorized access to site	Staff turnover and absence
Legislation: employment/ health & safety/ environment/ other	Welfare facilities limited Health concerns re light/ noise levels High staff turnover	Need to substitute safer materials Use of PPE Health surveillance in paint shop Maintenance Disposal of hazardous waste	Worker protection measures Need to up-date H&S training No proper consultation system in place Protection of young workers Noise levels Fire hazards	Noise levels Hazardous waste and scrap Training and supervision
Security:	IT systems not secure Authorized access systems not fully working	Data protection Need to monitor internet use Virus control Secure storage of vehicles	Potential arson at edge of site Unauthorized use of vehicles	Vehicles IT systems
Competition: industry/ consumer/internal	Competition for skilled staff Retaining trained staff Few young people in industry	Compliance with quality system in workshop Need to replace 2 machines Speed of finishing goods Stock control	Image of premises Protecting access to supplier base Competing for skilled staff	Competing for staff Quality of product
Finance: internal/external	Cost of training Rising cost of employment High sickness absence over 12 months period	Cost of waste/scrap Cost of rework for low quality Increase in customer returns	Cost of replacing old equipment Need noise reduction measures Cost supply of PPE for new staff Increasing wage bill Access to pensions 2001 onwards	Employment costs Investment in plant & machinery

Table 3.3 Risk factors identified: group (c) processes/performance

Risk factors	*Processes*	*Performance*	*Common features*
Employment: location/skills/ demographics	Lack skills in new workers Little choice about process Some work repetitive	Targets not set consistently across firm Data collected incompatible across firm Discipline procedures too heavy-handed	Way work is measured against targets Skills
Legislation: employment/health & safety/environment/other	Need to replace some equipment with safer versions Potential health problems in paint shop Manual handling Safe disposal waste materials Fire hazard Noise levels	Internal procedures for dismissal not used correctly RIDDOR details not collated No-one monitoring sickness absence Need to review power use re Climate Levy	Replacing equipment Use of existing systems
Security:	High damage levels to portable equipment Use of trucks on-site Power supply and computer-aided processes	Close-knit local community and access to sensitive information	
Competition: industry/ consumer/internal	Becoming out-of-date Need to use new materials Transport costs	Inconsistent against MSS criteria Poor industry bench mark rating	Rating against internal and external criteria
Finance: internal/external	Insurance costs Costs of rework Cost of sales Stock control and cashflow	Cost of absence Potential cost of Levy Potential loss of market share	Cost of poor quality

Table 3.4 Risk factors identified: group (d) planning/policy

Risk factors	*Planning*	*Policy*	*Common features*
Employment: location/skills/ demographics	Access to temporary staff Involving mobile staff Little current knowledge of future skills needs Changes needed to employment practice	Existing employment policies not adhered to No policy on disability or rehabilitation Discipline & grievance procedure too cumbersome	Employment practices Identify future needs
Legislation: employment/health & safety/environment/other	Use of new materials in planned production Time available to existing staff to find new suppliers Lack internal competence on H&S	Need for smoking policy Gaps in compliance with H&S regulations Policy on disposal of obsolete goods	Urgent need to ensure compliance with legislation
Security:	Poor base of relevant data to draw on	Security of personnel records	Access to data
Competition: industry/ consumer/internal	Access to relevant market data, internal and external Technology expertise	Need for eco-friendly purchasing Market awareness	Market analysis
Finance: internal/external	Minimal value remaining in existing plant Substantial investment required, both short- and long-term	Confirm policy on Late Payment Returns needed on capital invested	Investment

3.2 Who is likely to be affected?

Having completed this stage, a further column, Column 6, could be added if preferred in order to identify which individuals or groups of people are most likely to be affected by the potential risks specified. It is important to include reference to:

- individual workers at all levels within the organization, including those on part-time or temporary contracts; shift workers; home or mobile workers; people not based permanently at just one site; drivers and transport workers; agency staff
- other workers on-site not necessarily employed by the firm directly; other commercial occupiers of premises
- customers or clients, both on- and off-site; suppliers
- shareholders and other stakeholders in the organization; directors and senior management staff
- investors; finance providers such as banks; insurance providers
- visitors on-site, whether authorized or not
- residents, industrial and commercial users in the vicinity; local wildlife habitats and the environment.

This may be clearly restricted to a specified location, business unit or division.

> Example: Storage facilities at a particular plant are clearly inadequate for the growth in production activities taking place there, leading to increased exposure to hazards associated with unsafe racking or movement of goods; increased waiting times for unloading deliveries; wasted time due to ineffective recording/stock control systems; drop in quality compliance levels.

In other situations, particularly health and safety or employment issues, there will be individuals who are most at risk.

> Example: The sales force relies on vehicles being safe and roadworthy, fuel readily available and roads being passable. Fuel shortages, major road works and flooding increase the stress levels among these workers and greater incidence of 'road rage', plus longer working hours due to hold-ups.

Purchasing or performance factors are likely to impact throughout the organization, especially where business units are widely spread geographically and there is a greater degree of central control than devolved autonomy in such decision-making.

There are two more elements that need to be considered in order to evaluate fully the potential risks to the business, that is the level or severity of harm that is likely to occur and the likelihood that it will happen. There are many different ways to approach this stage and Chapter 4 looks at some of them.

Chapter 4

Evaluating the risks

4.1 Rating the extent of potential harm

No matter how detailed the work in the previous section, it is still primarily a subjective exercise to allocate some form of rating to all the different elements that constitute a 'risk' to the business. In addition, those who want to see that such an evaluation has been carried out will come with their own set of objective measures against which to validate an existing version. It is impossible, therefore, to suggest there is just one correct method. Having said that, it is valid to produce an evaluation of risks based on a comprehensive base of knowledge about the potential hazards or risk factors, knowledge of the context in which the business operates and a rating system that is as simple or complex as it needs to be. It may be easier to allocate a numerical value to these evaluations, or to make a judgement based on criteria such as high/medium/low.

In a health and safety context, the possible severity of harm associated with the hazards identified is generally fairly straightforward. There is a considerable body of information available to illustrate the type and severity of harm likely to occur from exposure to specified hazards. For example, physical injury to upper limbs is associated with the use of machines used in wood-working shops, plus the development of nose/throat/lung cancer from exposure to carcinogenic hard-wood dust particles. In this situation, identification of individuals most likely to be exposed to such hazards is fairly obvious, as is the location.

The extent of harm associated with exposure to health-related risks is sometimes more difficult to establish, particularly those with long latency periods between exposure and appearance of symptoms, for example some of the chemicals used in metal manufacturing processes during the 1950s and 1960s. Previous use of substances or materials may be an additional factor that needs to be considered against 'product' or 'process' in earlier chapters, especially where such substances have subsequently been reassessed as hazardous.

In the context of health and safety, typical ratings for potential harm include:

1 'Slightly harmful' or 'low' rating – such as minor or superficial injuries that might or might not require first aid treatment; levels of noise or other emissions at current minimum levels allowed by law
2 'Harmful' or 'medium' rating – such as serious sprains or fractures, burns, concussion, that result in lost time or require a hospital visit; potential for harm to some vulnerable groups of workers
3 'Extremely harmful' or 'high' rating – including potential for major injuries, fractures, irreversible chemical damage; significant hearing damage; and,of course, death.

Fire or environmental risk factors will need to be considered differently given the type of harm or damage likely to occur. The severity of harm or damage is not so easily related to individuals carrying out specific tasks, but is often more invasive or all-inclusive. There are similarities between the way a fire might spread and a chemical spillage, for instance, where the impact can be felt over a fairly wide area. Typical ratings in this case could be:

1 small scale, slow-release localized – can probably be tackled safely in the early stages by a competent person
2 small scale, localized to start with but potential for rapid spread (either fire or spillage)
3 likely rapid spread, and/or potential for producing toxic fumes or smoke affecting a wide area
4 instant spread possible over wide area, perhaps via a spread of chemicals, combustible materials or dusts
5 significant potential for explosion.

In these situations, the impact may well be on others outside the organization, rapid and fairly localized, slow and widespread, or indeed spectacularly widespread in the event of explosion. If preferred, the ratings could be combined as:

(a) 1 and 2 equivalent to the 'low' rating
(b) 3 equivalent to 'medium' rating
(c) 4 and 5 equivalent to 'high' rating.

Severity of harm likely to arise from the security hazards identified will depend to a large extent on the size and range of business activities undertaken, as well as the industry sector. The level of sensitivity of data held will determine the extent of harm likely if it is lost. The loss of a few hundred pounds from the till of a local shop will have a much greater impact on the business than the same amount lost through staff pilfering in a multinational organization. However, such losses can have a wide-reaching impact on other elements of risk facing the firm, certainly financial and competitive risks. Ratings might include:

1 'Low' rating – inconvenient loss, little impact on other elements of the business, already 'costed in'
2 'Medium' rating – not immediately or easily recoverable, not fully covered by insurance; impact on other elements of the business; some public embarrassment
3 'High' rating – severe impact on financial viability of business; far-reaching impact of bad publicity; non-recoverable loss.

Competitive or financial risk factors are often less easy to quantify for potential severity of harm likely. This may be due to the potential for misinformation or incomplete data on which to base analysis, the need for optimism (whether well founded or not!) and the impact of events outside the direct control of the organization. Of course, management theory has developed to take these points into account and various financial modelling and theoretical tools are available to assist in this. At corporate level within large organizations, systems will already be in place to evaluate these risks, possibly more so than for operational risks. However, some readers will not have access to such facilities, so a similar approach to previous risk ratings for potential severity of harm is included here.

Global trading and the increased commercial use of the internet may not necessarily represent a significant problem, as many industries rely on a fairly local customer base in regular face-to-face service provision. On the other hand, the firm may be able to tap into a much wider customer base than was previously the case. Consumer pressure may be a more significant factor, as will increasingly powerful single-interest lobbying groups targeting specific industries for high-profile action. If larger clients insist on all suppliers obtaining third party certification to demonstrate compliance with specified standards, this might attract a 'medium' or 'high' rating against severity if the cost of such certification becomes prohibitive. Ratings might include:

1 'Low' rating – slow change of fashion trends for product; established firm with few major competitors locally; relevant standards already part of normal operations
2 'Medium' rating – increased compliance requirements from Trading Standards or other enforcement bodies leading to additional investment; external factors such as fuel shortages or high fuel price increases leading to significant disruption to delivery mechanisms, staff travel – more severe impact in some rural locations
3 'High' rating – traditional industry in decline; major employers leaving region leaving local support industries in critical position; lobby groups severely restricting ability to operate or attract necessary funds (such as animal laboratory testing facilities).

More so than in other areas, financial harm will depend entirely on the way the business is structured, funded and organized. The severity of potential harm relates to cost of operations relative to sales; ratio of borrowing to value within the business – an issue no matter what size the firm; cost of borrowing, access to funding when needed, interest and exchange rates outside the firm's control; the increasing cost of insurance cover coupled with the reduction in level of cover provided. These elements are compounded in some instances by the expectations of shareholders, perceptions of financial markets and potential loss of confidence by other stakeholders. The relevant ratings of 'low', 'medium' and 'high' can be used to identify how severe the potential impact will be on the organization, from having to economize at the lower end to consideration of divesting part of the organization or selling up altogether, or indeed insolvency. As with other factors, such as legislative issues, the potential for fines or penalties being levied may be a factor to include in the evaluations.

The immediacy of impact on the business could well have been an integral element of the severity of harm assessed under each of the factor headings, but if not it warrants specific consideration. With some risks, the fact that it might be 2 or 3 years before the impact is felt adds a further dimension to the evaluation, even more so if the time scale is further into the future. A separate, but just as significant, issue to include in the calculation is whether the potential impact will be represented by a short-term 'blip' in operations, have ongoing long-term effects on the way the firm performs, or indeed requires a fundamental change in structure or approach. Many issues seen as specific to a few individuals can have far-reaching effects on other workers or outside observers.

Although more thought is required to allocate a number score rather than a low-to-high rating, it does provide the opportunity to consider sliding scales for severity of potential impact, immediacy, short- and long-term effects and a spreadsheet basis for comparing the findings later. A useful scoring system is one based on either a maximum 30 score, or perhaps 50 score, allocated on the following basis:

1. Low rating – up to 10 score
2. Medium rating – between 11 and 20 (or 11 and 40)
3. High rating – between 21 and 30 (or 41 and 50).

4.2 Evaluating the likelihood that harm will occur

Given the now very comprehensive picture of potential hazards to the business and no doubt an extremely depressing picture of everything that could possibly go wrong, the next stage is to assess the likelihood that this harm, injury or damage will occur. Clearly, there are already many forms of control in place to ensure the potential damage is not so severe, whether these are physical controls or guards, safe working practices and systems properly supervised and monitored, contingency plans in the event of disruption to usual transport routes, or other valid means to reduce potential harm. Some of these controls will be considered in Chapter 5, but for now the 'likelihood' factor will need to be assessed against some relevant criteria.

For instance, the more people exposed to a particular hazard and the longer the exposure time, the greater likelihood that harm or injury will occur. This might be due to the compounding effects of exposure over time, such as harmful substances acting as sensitizers; complacency and carelessness as activities become a familiar habit; increased exposure to high-risk activities such as driving where the individual is not necessarily in sole control of events.

Example: A 45-year-old owner of a sawmill had part of his foot severed by a bandsaw, despite working in the plant for 25 years. He noted 'It's easy to become a bit blasé about working practices after so long in the job'.

On the other hand, there may be a greater likelihood of harm due to inexperience of the individual, lack of knowledge and awareness, or forgetfulness if the activity only occurs infrequently. Gradual breakdown or wear-and-tear of equipment/machinery/structural features/infrastructure will increase likelihood that harm or damage will occur, as will periods of close-down for repair and maintenance.

However subjective the evaluation might be, it should now be possible to produce a reasonable picture of:

1 hazardous conditions/properties/processes that could potentially cause harm, injury or damage
2 what this harm, injury or damage might be; who could be affected; and how serious the result of exposure might be
3 the likelihood that such harm, injury or damage will occur, taking into account any control measures that exist.

Tables 5.1 and 5.2 in the next chapter illustrate how the potential harm and likelihood factors support the decisions on priorities for action.

Chapter 5

Controlling the risks

5.1 Control measures

Various ways of controlling risks have been mentioned in passing already and, clearly, there are as many ways to control or reduce the impact of a hazard as there are types of hazard. Different interpretations of the term 'control' are also possible. In this instance we refer to a control measure as:

> an action/device/strategy intended to eliminate/alleviate/reduce the negative impact on the business or individual of a situation or event.

Readers are likely to be familiar with most controls commonly used, although some may only have direct experience in the context of health and safety controls, or of financial controls. It is therefore worth reviewing the different categories of controls that could be employed in an individual firm.

The following headings are used for convenience, some things falling clearly into one category or another, others included in a category on a fairly arbitrary basis.

1 Physical controls
  - (i) Health and safety – wide range of physical controls such as guards, rails, barriers; lifting and handling equipment; fail-safe systems and stop buttons; personal protective equipment (PPE) such as hats, goggles, boots, gloves, masks, harnesses, breathing apparatus; ventilation and exhaust systems; vehicle alarms and specified routes on-site
  - (ii) Fire – fire, heat or smoke alarms; notices and warning signs; fire fighting equipment; circuit breakers; restricted access areas or containers for flammable materials; fire doors; sprinkler systems
  - (iii) Washing and decontamination facilities; tachometers; clocking in records
  - (iv) Use of CCTV and security badges; time-activated locks; access barriers
  - (v) Security systems for access and use of IT equipment
  - (vi) Relevant insurance cover.

2 Behavioural controls
   (i) Relevant training provided at all levels of the organization; specific qualifications required to operate in some areas
   (ii) Communication systems that reach all sectors within the organization; regular meetings to provide up-dates on progress
   (iii) Individual responsibilities and boundaries of authority clearly identified and agreed; provision of adequate supervision
   (iv) Incentives, reward schemes and other methods used to motivate and encourage staff; internal promotion and advancement possibilities
   (v) Culture; attitudes towards taking rest breaks and holidays when due.

3 Organizational or procedural controls
   (i) Planned maintenance programmes; regular checks on equipment and machinery as the norm; safe working procedures and 'Permit to Work' systems in place and adhered to
   (ii) Regular health surveillance where necessary; hearing and sight tests provided; review of work patterns to reduce repetitive movements
   (iii) Emergency procedures in place and tested regularly; contingency plans in place (for spillages, for example); washing and decontamination procedures adequate for type of work carried out
   (iv) Consultation with employees via their union or worker representatives
   (v) Use of security passes and authorization arrangements complied with; culture of enforcing internal rules (such as wearing hard hats or hearing defenders) as the norm
   (vi) Adequate recruitment policies and practices; compliance with equality of opportunity legislation; staff screening systems; employment contract agreements
   (vii) Adequate allocation of resources
   (viii) Signatory procedures in place for transactions; log on and password procedures; systems to safeguard lone workers
   (ix) Use of Management System Standards such as ISO 9000 or ISO 14000 series, or BS8800 (for OH&S); use of industry standards; use of Approved Codes of Practice (AcoPs)
   (x) Collection of data at regular intervals; senior management/board level commitment to protecting people from risks.

The list is not exhaustive, but highlights how difficult it is to categorize elements that overlap and interact with each other so closely. There may be other actions, devices or strategies that act to reduce the potential impact of risks identified in any one organization, or a combination of measures that are necessarily much narrower than those listed here. In any event, control measures that exist can be considered alongside the risk factors identified earlier, either within the framework of the five risk factors:

- Employment
- Legislation

- Security
- Competitive
- Financial

Or against each of the ten headings used for the 10 Ps:

- Premises
- Product
- Purchasing
- People
- Procedures
- Protection
- Process
- Performance
- Planning
- Policy

Having considered this substantial amount of data collected, there is room for a critical look at the validity of findings so far to 'make judgements about adequacy of controls in place and identify gaps in provision' (Chapter 1), before deciding priorities for future actions that might be needed to correct the situation.

5.2 Systems of control

As businesses grow it is inevitable that more formal systems than those generally associated with smaller firms will be needed. Experience suggests that around 50 employees is the point at which the entrepreneurial, flexible management approach no longer works effectively (on the assumption that it ever did). When carrying out such a far-reaching analysis as this one, it is important to confirm that the data it is based on are relevant, accurate and up-to-date. Even more important is the need to be critical about whether the existing controls are in place in reality, or are assumed to be working because the rule book said they should be.

The question posed is, therefore:

> How effective are these controls and what systems are in place to ensure they are appropriate, working, adhered to by everyone and still effective?

It may be useful to use the following summary as a prompt to consider this question more closely.

Employment controls

- Recruitment: are job and personnel specifications relevant to the jobs as they currently are, rather than how they used to be? Are specifications appropriate for bringing in the correct skills and expertise required now and in the future? Are future skills identified in plenty of time in order to develop them in-house or recruit from outside? Who makes these decisions and on what basis? What flexibility of working patterns is available to workers?

- Equal opportunities: does the mix of workers reflect the make-up of the local community? Is there a mix of ethnic backgrounds among the workforce/a mix of age groups/a fair balance of male and female workers in all job areas? Have efforts been made to accommodate the needs of disabled applicants or workers?
- Training: is training provided to workers across the organization? Are there restrictions on access to some training based on age or sex of the worker? Is this based on valid assumptions? Is there a full programme of induction training, including when people move from one site or division to another within the same company? What is the balance between internal/external/general/industry-specific training? What support do you provide for workers who wish to undertake their own training or studies?
- Communication: are adequate systems in place to ensure proper channels of communication are maintained throughout the organization? Is there real commitment to proper consultation with workers? Are all groups of workers involved, including those on shifts or regularly working off-site?
- Supervision: is direct supervision appropriate, effective and maintained? Have vulnerable groups of workers been identified for additional support if necessary? Are supervisors and managers trained adequately in this role, rather than just their area of technical expertise?
- Premises: how often are facilities reviewed to confirm they are still adequate? What is the refurbishment programme time scale? Who decides on structural alterations or decorating designs?

Legislative controls

- Compliance: how does the organization stay up-to-date with legislative changes? How are people notified of changes and how they will be affected? Is specialist advice available internally/externally/cross division? How is this monitored? Are legislative differences worldwide fully taken into account? Is the Lead Authority Partnership Scheme (LAPS) used for cross-boundary agreements in UK?
- Health and safety: how are results of risk assessments recorded? Is this procedure standardized across all divisions? Have people received training in carrying out risk assessments? What are internal procedures for reviewing and up-dating risk assessments? Are these carried out when new plant or machinery is introduced? Have vulnerable groups of workers been identified?
- Employment law: is the internal discipline and grievance procedure effective? Has the list of 'gross misconduct' offences (often referred to as dismissible offences) grown so long as to include every possible minor infringement, affecting morale and motivation? Are provisions under maternity leave, working time and other major legislation applied equally throughout the firm? Is there a history of cases being brought against the firm in tribunals, civil or criminal courts?

- Environment: are waste management systems effective? Are emissions checked as often as necessary? Are maintenance programmes established and adhered to? Are all relevant licences obtained and conditions complied with?
- Records: are all relevant records complete and up-to-date? Is responsibility for completing records correctly allocated? Are records accessible or restricted appropriately? Are the data collected relevant and in a usable format? Is information passed to the relevant authorities according to legislative requirements?

Security controls

- Physical security measures: are insurance providers satisfied that adequate security measures are in place; what is the record of claims, break-ins, damage to property? Are security providers vetted sufficiently? Are systems working correctly?
- Access: is everyone provided with correct authorization? Are new staff screened before they arrive (for example through police or social services records) if relevant to their position? How are computerized records protected? When was this system of protection up-dated and by whom? What checks are in place to confirm the protection works effectively? How are secure data sources accessed when authorized personnel are absent? Are systems in place to combat viruses or deliberate sabotage of data?

Competitive controls

- Information: how are customer complaints or returns dealt with? Have complaints or reported faults risen or declined in recent years? What trends have emerged and what reasons have been identified for them? What communication channels exist across and between companies?
- Marketing: how are internal and external market intelligence data used and at what levels in the firm? How are they collated and disseminated? What measures are used to monitor the results of market strategies? How and when are these reviewed and amended if necessary? How have ratios of marketing costs to sales or profits altered over last 5 years and why? How effectively is the internet used for commercial transactions, if at all?
- Standards: how does the firm compare with other industry players in bench marking exercises? Are relevant industry standards used effectively across part or all of the organization? How are changes to these requirements implemented? Are the principles of continuous improvement adhered to across the whole organization? If not, where are the gaps and are they acceptable? Is the application of ISO 9000 and/or ISO 14000 MSS requirements fully complied with, maintained adequately, supported by organizational commitment? Have these systems been fully integrated with each other and any others that are relevant? Have differences world-wide been taken into account without putting some nationals at a disadvantage compared with others?

Financial controls

- Procedures: do accounting procedures fully comply with legislative requirements? Are the data generated available to the right people at usable intervals for them to make financial decisions, weekly/monthly/quarterly rather than after the end of the year? Are financial records fully compatible across the organization, whether based inside or outside the home country? Are records fully up-to-date at any given time in the financial year?
- Cash flow: is cash flow given sufficient attention as well as sales/costs/end of year profits etc.? Are late payment procedures in place and working adequately? Does the firm also ensure it pays bills on the same principles of late payment to avoid restricted or delayed supplies? Is factoring used for collecting bad debts? What is the debt recovery pattern over the last 2 years?
- Insurance: when was the last time insurance cover was renegotiated rather than just renewed when due? Is insurance cover adequate for new or different risks that may have arisen, such as new plant or machinery to replace the old, or exports to different countries from those cited in original proposal forms to insurers? Has value of insurance cover kept pace with increases in value of property? Is compulsory Employers Liability cover held? What is the claims record – any patterns emerging?
- Capital: how are changes to exchange rates/bank rates/inflation/borrowing terms/VAT monitored? How are stock market movements monitored and taken into account? Are investment reviews and evaluations carried out internally/externally/mixture of both? How often and is this satisfactory to ensure the best for the firm? Are compliance costs, penalties etc. identified and taken into account when financial planning is undertaken? Is access to technical and financial expertise available?

By this stage, none of these points will necessarily be different from what has already gone before, but referring back to them in each context reinforces their importance and ensures they are not forgotten. Although adding still more dimensions to consider when assessing the risks to the business, they should also help to clarify the elements of the business that may have been neglected or perhaps under-rated previously and emphasize the need for a holistic approach. If the organization is complex and a numerical rating is applied at various stages of the process, a clearer grading system should emerge so making the next stage of deciding priorities for action that bit easier.

5.3 Deciding priorities for action

Alongside the ratings given to risk factors are still further questions that need to be addressed. Having assessed the extent of harm or damage likely and potential disruption to business activities, the organization's ability to recover from the impact will be crucial to the prioritizing process. A critical look at the robustness of internal systems and evaluation of their strengths and weaknesses within all areas is a vital element of the subsequent decision-making process. If a wide range of risks needs to be addressed, then some form of prioritizing has to take place in order to ensure action is taken, within a given time frame, to eliminate the risks or reduce them to an 'acceptable' level. Less demoralizing than identifying lots of areas that need to be addressed is for nothing to be done at all to correct them.

Listing factors in descending order of scores applied makes it easier to start the prioritizing process and to illustrate how wide-ranging and complex the risks to the business are. However, the danger is that just those at the top of the list are tackled, leaving the lower-scored issues to be dealt with on a more ad-hoc basis.

If the original risk factors identified are collated with recurring or overlapping themes combined, a list of 'primary risk factors' will emerge. These can then be plotted on the grid suggested in the example (Table 5.1) and should result in a scattered response pattern.

In this example, three divisions are used on each axis, resulting in nine categories ranging from:

- a 'trivial' or low risk score of **(1)** representing unlikely or low-harm events
- a score of **(2)** where it is quite likely that an event will occur given the factors identified, although the resultant harm may not be great
- a score of **(2)** where an event is unlikely, but if it did occur the harm or damage could be significant
- medium score of **(3)** where an event is very likely, but resultant harm is considered 'trivial'
- medium score of **(3)** where an event is likely to occur and resultant harm or damage could be significant
- medium score of **(3)** where an event is very unlikely, but if it did occur the result could be extremely harmful

Table 5.1 Assessing the risks – three categories

	Slightly harmful or low-level harm	*Harmful*	*Extremely harmful*
Low likelihood/ highly unlikely	TRIVIAL RISK 1 *******	**** 2	*** 3
Medium likelihood/ likely	******* 2 **********	** 3	4
High likelihood/ very likely	3 ***	************** 4	INTOLERABLE RISK 5 *******

*** = risk factors identified

- a high score of **(4)** where an event is quite likely to occur and the results extremely harmful
- a high score of **(4)** where an event is very likely to occur and significant harm or damage is likely
- an 'intolerable' or high risk score of **(5)** where an event is very likely to occur and would be extremely harmful if it did – representing the need for urgent action.

Immediate action is clearly required for those risk factors that appear in the last category above – that is with a score of **(5)** – in that the potential damage could be devastating to the individual or business and is indeed very likely to happen. These factors require remedial action within the short term, both to reduce the likelihood that they will occur as well as increasing protection and reducing the potential impact if it does. Risks that fall within the boxes showing a score of **(4)** also represent significant risks that may require urgent and serious consideration. Ultimately, the aim is to generate a significant shift of risk factor scores from the bottom right-hand box as close as possible to the top left-hand box.

On the assumption that levels of risk now identified are a fair reflection of what is currently happening in the business, a decision also has to be made about which risks are deemed to be 'tolerable' and need no further action at this stage. A very low rating score does not mean the risk factor is immaterial. After all, the scoring system has been applied to risk factors evident in the firm, not a hypothetical list of factors that might or might

Table 5.2 Assessing the risks – five categories

	Slow spread/ localized	*Rapid impact/ localized*	*Harmful*	*Rapid impact/ wide damage*	*Explosion/ severe harm*
Few people affected	1 * * *	1 * * * * * *	2	2	3 * *
Highly unlikely	1 * *	1/2	2 * *	3 * *	4
Likely	1/2	2 * * *	3 * * * *	4 * * * * *	4
Very likely	2	3	4 * * * *	4 * * *	INTOLERABLE 5 * *
Many people affected	3	4 * *	4	INTOLERABLE 5	INTOLERABLE 5 * *

not be present. Indeed, a large base of 'trivial' risks can in themselves increase the rating for other factors, such as motivation of staff, or quality of production and must therefore be seen to be dealt with effectively. As noted previously, this is a subjective analysis to some extent and will reflect the attitudes and beliefs of those providing the assessment to a greater or lesser degree, but nevertheless it provides a firm base to consider both short- and long-term changes that may be required to manage risks effectively.

The second example (Table 5.2) shows a more extended table based on five-by-five categories, allowing for a finer delineation of elements that impact on the final position in the chart. Again, the bottom right-hand corner represents the most significant risks to the business, this time shown as three shaded boxes rather than just one.

If a scoring system of 1–10 is used to evaluate the potential impact of each risk factor, the emerging pattern when presented in the further example (Figure 5.1) will illustrate the highs and lows of the range. Decisions have to be made about the balance between levels of scores/ urgency of action required/order of priority for action and although a

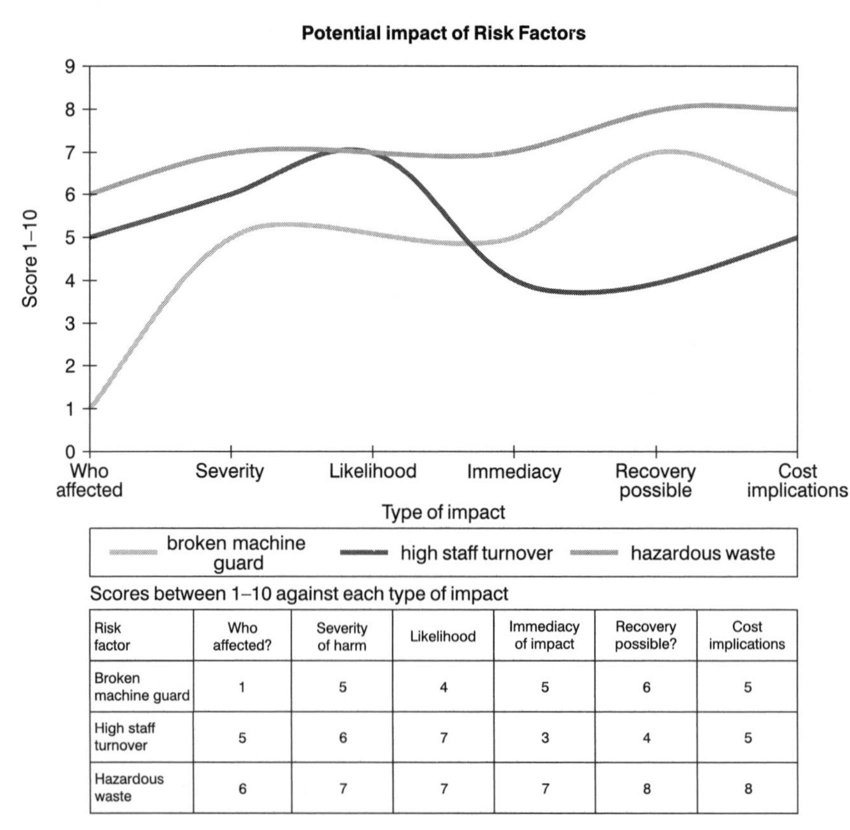

Scores between 1–10 against each type of impact

Risk factor	Who affected?	Severity of harm	Likelihood	Immediacy of impact	Recovery possible?	Cost implications
Broken machine guard	1	5	4	5	6	5
High staff turnover	5	6	7	3	4	5
Hazardous waste	6	7	7	7	8	8

Figure 5.1 Assessing the risks – using a numerical score

mid-range 'priority' line may appear as a natural divide based on the evaluation and scoring process used, it may need to be established and justified separately. So, for example, the scores for potential severity of impact and likelihood that an event may occur represent the significance of risk to the business that is compounded by the firm's ability to recover from such loss.

Finally, cost implications of not taking action have to be balanced against likely cost implications of dealing with the risk factors. In this case, it must be stressed that cost may indeed be a factor in the level of action taken to reduce risks, but must not be the basis for a decision to take no action at all – certainly in the area of legislative risks. Over-reliance on comparative measures for deciding priorities may also conceal the fact that a fairly low total score incorporates one element with a very high score for severity to an individual, where action must be taken in order to protect the individual from harm.

Summary of the prioritizing process

1. *All* risks identified are collated to reduce overlap and repetition and to establish a list of primary risk factors
2. Risk factors identified are listed in descending order according to total score allocated
3. Plot scores on grid or spreadsheet
4. Decide criteria for consideration of urgency of actions required
5. Identify where urgent action is required, such as clusters of risk factors found in bottom right-hand shaded boxes of grids
6. Consider what actions can be taken quickly to alleviate the potential harm, whether low or high scoring factors
7. Identify long-term actions needed, within acceptable time scales that do not further jeopardize the protection of people, property and the business itself
8. Consider cost implications, ensuring that they are not used as an excuse to avoid taking action where it is clearly required to safeguard the future of the firm
9. Review risk factors and re-evaluate on the basis of potential results of planned actions
10. Consider residual risk and hazards that will still remain despite planned actions. Note which elements are inherently hazardous and can only be contained through the use of a range of control measures
11. Define 'acceptable' or 'tolerable' risk and consider the level at which the organization accepts such risks exist and that sufficient action has been taken 'as far as reasonably practicable' in the circumstances
12. Identify which factors can be reduced so as to virtually eliminate the perceived risks, for example through substitution of materials or changed processes and procedures
13. Ensure that any planned changes do not introduce further or new risks to the operation of the business.

Chapter 6

Case studies

To illustrate the way different types of business might apply the suggested approach to identify potential risks, four case studies are considered in this Chapter. There are two service-based industries that are fairly typical of today's provision. These are:

- Case study 1: health services such as doctors, dentists, veterinarians and other associated services
- Case study 2: call centres, either as independent units or as divisions of a larger organization.

There are also two production-based industries, which are of course a crucial part of UK industry. These are:

- Case study 3: food production and/or processing, whether as small-scale units or larger industrial plants
- Case study 4: engineering and manufacture, becoming more specialized and often focused in certain geographic areas.

The potential risks to each of these types of business are considered against the 10 Ps, risk ratings suggested against each of these elements and a chart drawn up to highlight where the priorities for action appear.

6.1 Case study 1: health services

Initial thoughts about likely risk factors

These services rely primarily on face-to-face contact with the customer or client, so important issues likely to be related to process/procedures/people/protection closely followed by premises/performance (this last one being externally imposed in some cases). Generally serving a local community rather than a global one, although larger practices with more than one site may face greater risks regarding Local Authority requirements, or different client groups. Bad publicity associated with the medical profession serves to undermine credibility and creates wariness and concern among patients or clients. This is sometimes seen as loss of professional integrity and standing and a more significant willingness to consider litigation in some cases.

Risk factors: group (a) premises/product/purchasing

Issues identified include:

- Premises: status and due date for rent review; although structure sound, as a 'listed' building, alterations are difficult to make whether for legislative purposes (such as adding fire doors) or to improve access for patients; expansion is limited as inner-city site and car parking is extremely limited; problems associated with crime in inner city area are growing, needing better security systems to safeguard staff as well as premises; business rates risen significantly during last 12 months.
- Product: ageing population in area so upward trend in treatment and preventive care needed; need to access relevant, up-to-date information on medicines and treatments; high cost of some pharmaceutical products and increased pressure to restrict supply; access to 'alternative' medicine services; very limited access to occupational health services in region; customers expect to receive better level of treatment as private rather than NHS patient and the need to get the balance between these provisions right.
- Purchasing: high investment costs to maintain levels of equipment and services; business materials easily accessible, but limited suppliers for some specialist equipment keeping prices high; increased use of disposable consumables (such as syringes) in treatment of individuals, based on increased risks to staff.

Risk factors: group (b) people/procedures/protection

Issues identified include:

- People: high proportion of skilled practitioners approaching retirement age; lack of well-qualified people being recruited in area or choosing to stay in inner-city practice; need for regular refresher training but not enough time to take it up; high staff turnover at lower levels such as surgery assistant.
- Procedures: increased customer base making appointment system difficult to maintain; pressure from inadequate time available per patient; cost of non-attendance at appointments and issuing reminders; keeping records correct and up-to-date; maintaining integrity of individual client samples (especially when sent off-site for investigation); data protection; procedures for removing clients from list/ notifying them/answering queries; refusing treatment, for example if client is abusive or drunk; collection of fees (from NHS and individuals) and impact of delays; fire, health and safety procedures partially in place; evacuation procedures in emergency situations not well rehearsed.

- Protection: increasing levels of insurance cover and potential for litigation claims; exposure of staff to pathogens/diseases; ionizing rays from use of X-rays – use now being reduced on site, based on risk assessment findings required under the 1999 Ionising Radiations Regulations (IRR) and future need for authorization to use; manual handling problems and potential back injuries due to nature of work with people or animals; vibration and noise levels associated with use of some machines; need for safe disposal of 'sharps' and rising cost of this disposal in city; cost of provision/maintenance/laundry of protective clothing.

Risk factors: group (c) process/performance

Issues identified include:

- Process: direct contact with client and potential for 'getting it wrong'; reliance on information from client and individual records being correct; potential problems associated with administering drugs/oxygen/anaesthetics/X-rays, both as correct procedures to safeguard patient and to safeguard staff from unnecessary exposure; difficult to judge how long each consultation will take; provision of service and access for clients out-of-hours.
- Performance: increase in external pressure to identify performance measures and monitor success against these; performance measures include waiting times for appointments/waiting times for referrals (often outside our control)/complaints received from clients/sick absences of staff/training received and qualifications achieved.

Risk factors: group (d) planning/policy

Issues identified include:

- Planning: need to agree the balance between private and NHS provision; partners all independent but need to have input to decision-making process; how to accommodate increasing base of older clients, their increased need of services available and costs that incurs; long-term planning for provision associated with inner-city health problems (poor housing conditions/unemployment/less disposable income/inadequate diet); make better use of support staff to reduce administrative burden on practice partners; establishing performance monitoring system; need to plan for advice on IRR compliance and monitor exposure to radiation over calendar year.
- Policy: equal opportunities – assistants still tend to be mainly young women, so difficult to change the mix; worker profile means policy on maternity leave etc. particularly important; limited availability of alternative work if pregnant worker unable to carry out normal duties; disability discrimination – problems with old premises difficult to overcome as physical changes limited (some provision has been made) – particular issue given profile of clients; procedures needed to safeguard vulnerable sectors of patients plus vulnerable workers; lone

working (such as on-call visits) in some parts of catchment area, plus security of vehicles/drugs etc.; provision of physical intervention training to reduce potential harm to workers and patients; working time/breaks/early recognition of symptoms of stress.

Potential impact of risk factors identified

Table 6.1 and Figure 6.1 (a) compare a scoring system against five risk factors identified for this case study.

Table 6.1 Score 1–10 depending on potential level of negative impact

Risk factor	*Who affected*	*Severity of impact*	*Likelihood of impact*	*Immediacy*	*Recovery possible*	*Cost implications*
Listed building	3	3	5	5	3	5
Increasing demand on services	6	5	7	7	5	7
Administration procedures	8	7	8	7	6	8
New performance measures	3	4	6	5	7	7
OH&S issues	6	7	5	6	7	7

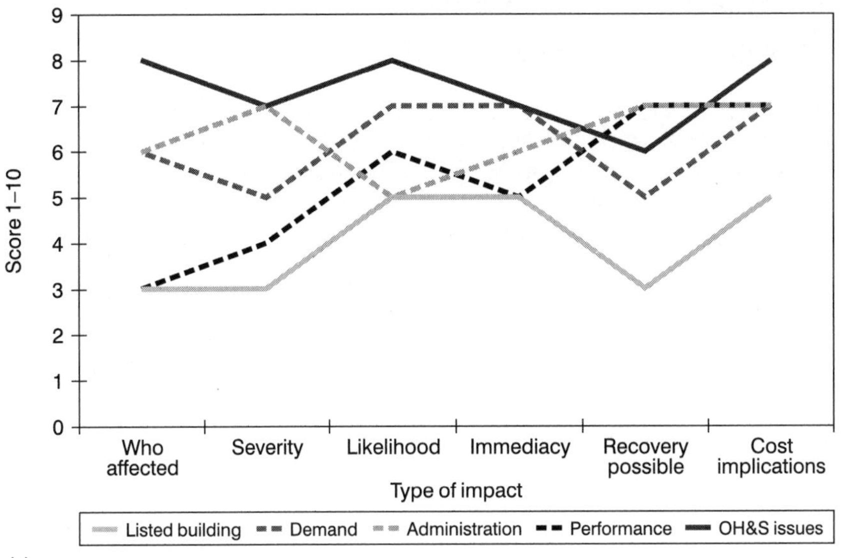

Figure 6.1 (a) Health services: potential impact of risk factors

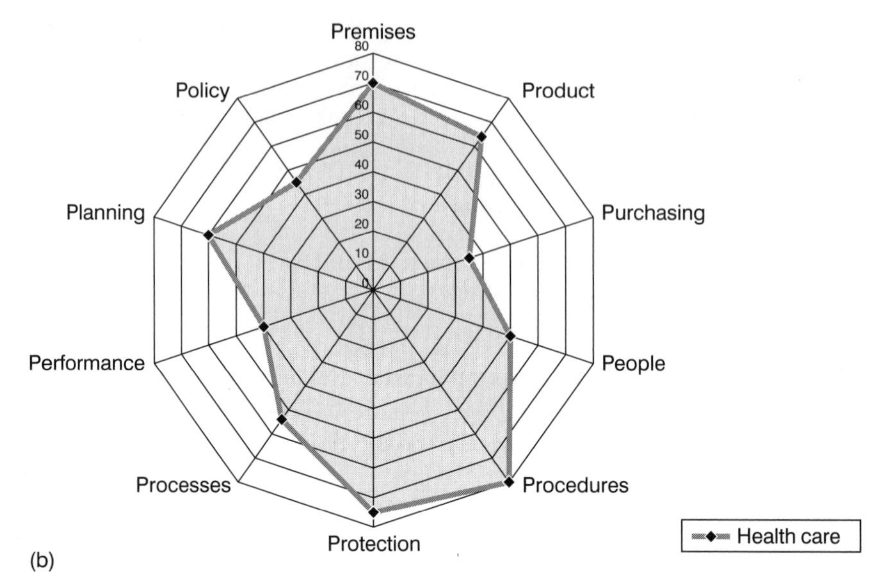

Figure 6.1 (b) Total scores

Figure 6.1 (b) compares the total scores between 1 and 100 allocated against each of the 10 Ps in order to identify potential priorities for the health service business featured in case study 1.

6.2 Case study 2: call centres

Initial thoughts about likely risk factors

A service providing direct interface between customer and supplier of products or services, though obviously at a distance – sometimes completely devolved to another region or country. Important issues likely to be related to the product/people/process, closely followed by premises/protection. The crucial factor is recognition that the philosophy of a call centre is based on providing a human interface with callers, rather than computer-generated responses, therefore staff are a *vital* component of the system. That is, after all, what the client is paying for – humans using technology to support this philosophy.

Risk factors: group (a) premises/product/purchasing

Issues identified include:

- Premises: location is irrelevant to location of client firms, or to customer as direct access not necessary, so can be in less expensive area of country; rural area has less satisfactory transport infrastructure in

place, making it expensive and/or difficult for workers to get there; rising fuel costs adding to this problem into the future; size and capacity of premises an issue, given the density of equipment needed and danger of becoming too cramped for comfortable working; work environment and conditions important to overcome perception of call centres being compared with the old industrial 'sweat shops'.
- Product: service is provided for a third party so removed from client; there is little choice on responses required by operatives, in some cases every word is specified; need to ensure sufficient knowledge of client's product or service; limited capacity available for responses per operative per shift.
- Purchasing: not initially considered a significant problem area, mainly related to access to training and other external facilities; equipment must be robust/adjustable (especially chairs, keyboard and screen)/suitable for intensive use/needs good user interface and screen image.

Risk factors: group (b) people/procedures/protection

Issues identified include:

- People: high staff turnover; recruitment difficult in some locations; skills needed including ability to speak clearly; training required for new staff, including use of equipment and adjusting to fit or reduce screen glare; very limited choice available to individuals in way they work.
- Procedures: use of display screen assessment procedures vital; ensure rest breaks taken; dealing with complaints or abusive calls from customers; need to take notice of staff complaints at early stage to reduce potential problems; escape procedures in event of fire or other emergency not clear.
- Protection: mainly health issues such as potential hearing loss and levels of background noise; static posture for long periods and potential for repetitive stress injury (RSI); rest breaks/pace of work/intensive use over long period; adequate light, heat and ventilation levels must be maintained; workspace ownership a potential issue.

Risk factors: group (c) process/performance

Issues identified include:

- Process: provision of information, advice and guidance on behalf of client; access to supporting information; convergence of telecommunications and computer facilities into single completely self-contained workstations; job design options limited as standard responses required; sitting in one place for long periods and associated stress factors for operatives.
- Performance: need to set appropriate targets in agreement with operatives; performance measures generally by volume and type of

calls/responses made/time taken/extra 'chat' with callers; volume of customer complaints per month; number of clients serviced and retained per annum; costs including wages plus staff turnover and absence levels.

Risk factors: group (d) planning/policy

Issues identified include:

- Planning: volume of calls and capacity ratios; number of clients serviced (whether in-house or external clients); shift patterns and ensuring adequate cover for staff absence; replacement and investment planning for equipment; return on investment.
- Policy: pricing agreements; payment terms agreed; collection of outstanding debts; wage and benefit agreements; equal opportunities across all sites and levels in firm; confirmation of commitment to taking staff complaints seriously and ensuring conditions are appropriate for the level of work expected.

Potential impact of risk factors identified

Table 6.2 and Figure 6.2 (a) compare a scoring system against five risk factors identified for this case study.

Figure 6.2 (b) compares the total scores between 1 and 100 allocated against each of the 10 Ps in order to identify potential priorities for the call centre featured in case study 2.

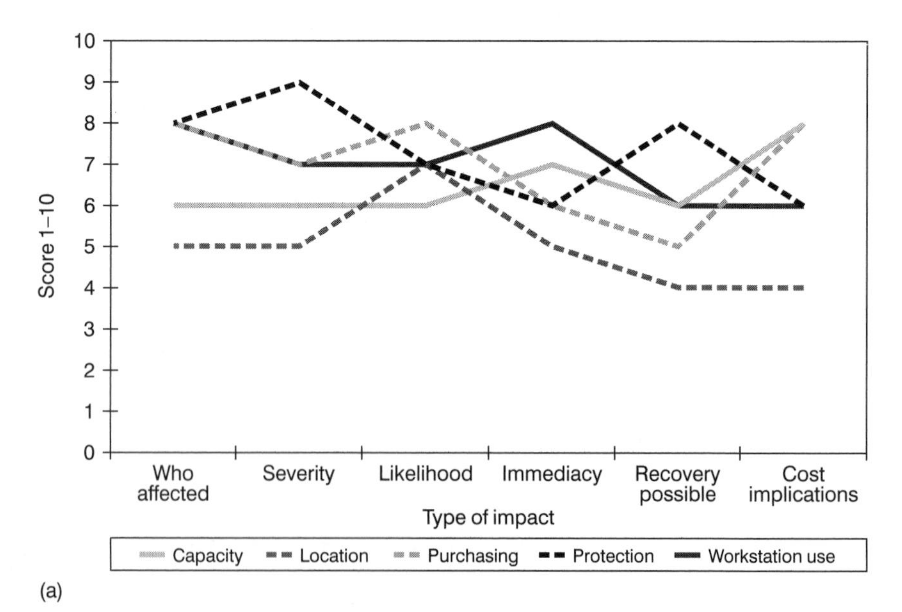

Figure 6.2 (a) Call centres: potential impact of risk factors

Table 6.2 Score 1–10 depending on potential level of negative impact

Risk factor	*Who affected*	*Severity of impact*	*Likelihood of impact*	*Immediacy*	*Recovery possible*	*Cost implications*
Capacity of premises	6	6	6	7	6	8
Location of premises	5	5	7	5	4	4
Purchasing right equipment	8	7	8	6	5	8
Protecting workers	8	9	7	6	8	6
Use of workstations	8	7	7	8	6	6

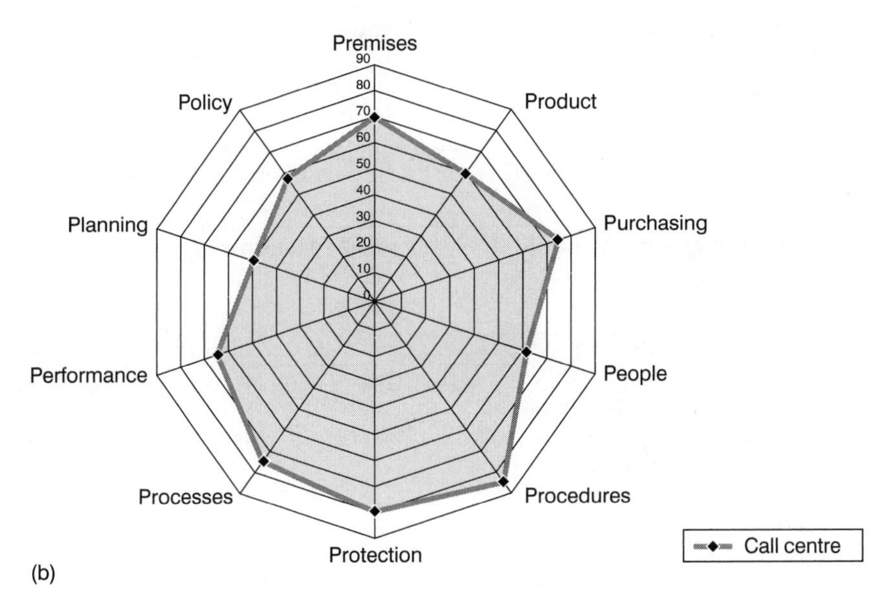

Figure 6.2 (b) Total scores

6.3 Case study 3: food production and processing

Initial thoughts about likely risk factors

With food production, a major concern is about delivery of consistent standards from a more variable base of foodstuffs that can be adversely affected at short notice. The industry is highly regulated, so legislative requirements are a prime issue and maintenance of adequate control systems. Issues are therefore purchasing/product/process closely followed by procedures/protection/premises. Throughout, people and planning are also important elements to consider.

Risk factors: group (a) premises/product/purchasing

Issues identified include:

- Premises: location and local amenities, particularly problem of unreliable power supply in certain extreme weather conditions; capacity of storage and distribution facilities on site; need for cold storage facilities and specialist delivery vehicles; safe storage of fleet on site and in transit; increasing fuel costs, but good local access to major transport routes.
- Product: labelling and packaging requirements considerable, with growing base of information needed to be included for consumers; conflicting scientific evidence makes it difficult to ensure consumer confidence (for example on GM components); need to include guidance on use as well as contents, to reduce potential for litigation claims; need to be able to change range quickly in response to national emergencies (such as BSE or foot and mouth crisis), as well as change in fashion trends.
- Purchasing: traceability of supplies; consistent access to correct standard of supplies; need to ensure ethical supply where possible; specification standards and tests for contamination; impact of seasonal/climate change on growing and production schedules.

Risk factors: group (b) people/procedures/protection

Issues identified include:

- People: high staff turnover at lower operative levels; limited skill base available locally as rural area rather than urban; some problems with literacy levels (non-English speaking workers); need to provide wide range of training internally; skills of drivers important given regular distances travelled for delivery.
- Procedures: usage of water during processing; levels and type of waste produced; waste storage/collection/disposal; potential leakage of contaminated substances/waste; monitoring driver hours and skills/training; evacuation procedures in event of emergency given variety of ethnic groups working on site; need to maintain quality and environmental management system standards fully.
- Protection: extremes of heat and cold in process; noise levels for some parts of process; need to maintain adequate ventilation and extraction systems; flow of dust and potential for explosion; protection of consumers with information on contents (allergies); quality assurance scheme for major customers; food and hygiene regulations strictly adhered to as well as H&S (use of HACCP); use of personal protective equipment where appropriate.

Risk factors: group (c) process/performance

Issues identified include:

- Process: high use of power so potential negative impact of climate change levy; significant levels of waste produced naturally; need to

maintain machines in constant use; use of high-pressure systems; combination of water and electricity in process so increased potential for electrocution; capacity of existing machinery and need to up-date with more computer-controlled, cleaner more efficient versions.
- Performance: constant need to ensure quality systems adhered to; volume and cost of scrap produced; amount of production down-time or delays in deliveries monitored; processing time and volume targets met monitored (recent problems with old machine awaiting replacement); staff/costs ratios; staff turnover rates and cost of recruitment.

Risk factors: group (d) planning/policy

Issues identified include:

- Planning: main problems associated with reliance on a major customer (Pareto's 80–20 balance) and risks associated with squeezing margins then cutting orders; increased use of e-commerce facilities planned; investment plans and expected returns for machinery replacement programme; contingency planning required in short-term to ensure supplies from new sources (given restrictions of foot and mouth outbreak in UK and elsewhere).
- Policy: recruitment based on non-discrimination and equality of opportunity; problems associated with identifying illegal immigrant applicants, given current rural location; minimum wage guidelines followed, although this will represent potential problem if levels raised significantly in future; cash flow and late payment policies in place, using factoring services where necessary.

Potential impact of risk factors identified

Table 6.3 and Figure 6.3 (a) compare a scoring system against five risk factors identified for this case study.

Figure 6.3 (b) compares the total scores between 1 and 100 allocated against each of the 10 Ps in order to identify potential priorities for the food production business featured in case study 3.

Table 6.3 Score 1–10 depending on potential level of negative impact

Risk factor	*Who affected*	*Severity of impact*	*Likelihood of impact*	*Immediacy*	*Recovery possible*	*Cost implications*
Labelling and guidance	5	5	6	5	5	8
Location of premises	7	8	6	5	7	8
Purchasing supply traceability	5	8	8	8	8	6
Power usage levels	6	7	5	4	7	8
Local skill base	7	8	7	6	7	7

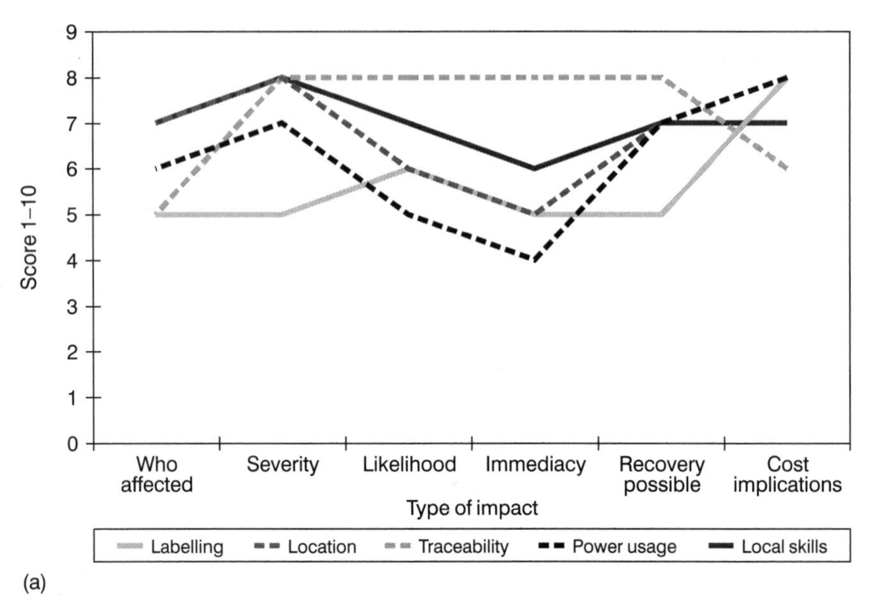

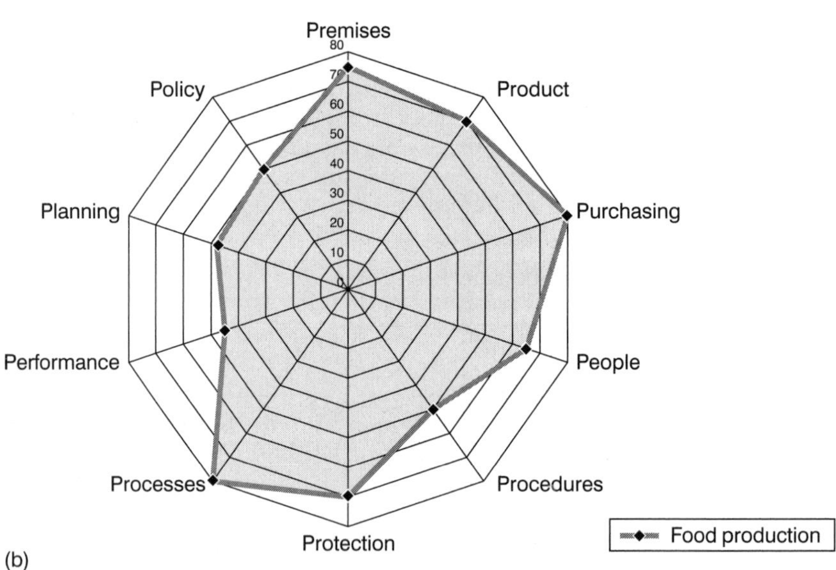

Figure 6.3 (a) Food production: potential impact of risk factors. (b) Total scores

6.4 Case study 4: engineering and manufacture

Initial thoughts about likely risk factors

As with the earlier case study on food processing, this is primarily concerned about consistency of supply to customers, although there is likely to be less variation in the quality of raw materials. Environmental and competitive issues are of particular concern in this context, as their potential negative impact can be considerable. The important factors are purchasing/product/process/procedures, closely followed by pemises/people/planning, although performance is of course a vital element too.

Risk factors: group (a) premises/product/purchasing

Issues identified include:

- Premises: capacity and layout; storage for large-scale raw materials/work-in-progress/finished goods; access for deliveries and vehicles used on-site; security systems for restricting access to visitors (especially those delivering goods); distance from main transport routes, particularly for exports and materials sourced from outside UK.
- Product: need for consistency of quality standards to customer specification; increased cost of packaging materials and regulatory requirements; minimizing use of packaging to only essential protection; major concern about producer responsibility for disposal of obsolete manufactured goods from customer; need mix of long-term and short-term design specification work to optimize use of machinery.
- Purchasing: increased cost and reduced supply of raw materials; use of internet to find new, renewable sources of materials for some work; need regularly to up-date computer-assisted machinery.

Risk factors: group (b) people/procedures/protection

Issues identified include:

- People: fewer tasks requiring non-skilled labour and more computer-literacy skills needed; local skill base reasonable, but training on specific machines still required; older average age of workforce – not a problem at present, but will become one in around five years time; need to bring manager-level skills in from outside for some parts of business.
- Procedures: emergency procedures for evacuation of plant need up-dating, particularly for potential hazardous chemical spillage and widespread impact on local community; safety procedures for use of heavy equipment; regular noise level monitoring in place; storage and handling of scrap, plus specialist disposal facilities used.
- Protection: traditional areas of health and safety protection still important, with increased emphasis on use of VDUs; security of design and manufacturing process information; need to ensure existing procedures are in place and working adequately (up-date risk assessments).

Risk factors: group (c) process/performance

Issues identified include:

- Process: capability of existing plant and equipment to meet current and future client specifications which are becoming more sophisticated; high use of power accompanied by need to maintain cleaner work environment, so replacement of ventilation system required.
- Performance: client expectations of adherence to quality and environment MSS; regular review for accreditation to ISO standards; individual performance targets set and monitored annually, up to senior management level; refresher training provided when possible, but in-house training expected to pass on relevant skills to production workers; returns on investment programme (currently year two of five) monitored; insurance claims and premium reviewed regularly to ensure adequate risk control systems in place.

Risk factors: group (d) planning/policy

Issues identified include:

- Planning: need to ensure adequate communication between planning/production/sales/purchasing departments to avoid last minute scheduling changes (and subsequent costs of scrap and changeover time); continued programme of replacing equipment; skills and recruitment planning needed to match planned changes to production profile.
- Policy: equal opportunities, but note success of recruiting in non-traditional areas still limited; need to consider rehabilitation policy to retain skills; confirm, in consultation with all interested parties, policy of emphasis on increasing high-specification products that are high-value, rather than high volume low-value production cycles.

Potential impact of risk factors identified

Table 6.4 and Figure 6.4 (a) compare a scoring system against five risk factors identified for this case study.

Table 6.4 Score 1–10 depending on potential level of negative impact

Risk factor	*Who affected*	*Severity of impact*	*Likelihood of impact*	*Immediacy*	*Recovery possible*	*Cost implications*
Use of safer materials	5	6	7	6	5	7
Quality failures	6	7	6	6	7	9
Disposal of hazardous waste	5	8	6	6	8	8
Security vehicles	6	8	7	7	6	9
Customer complaints	3	5	5	5	6	7

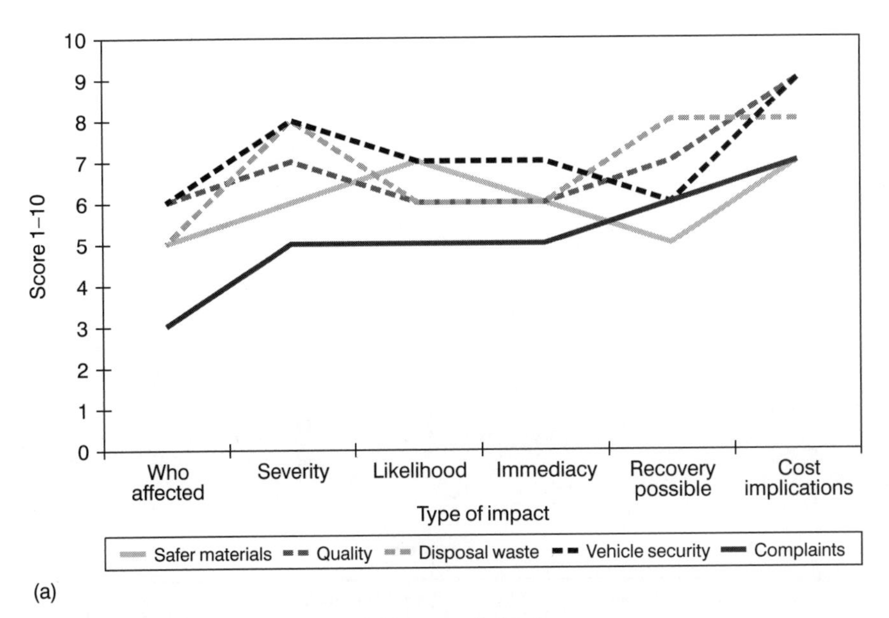

Figure 6.4 (a) Engineering: potential impact of risk factors

Figure 6.4 (b) compares the total scores between 1 and 100 allocated against each of the 10 Ps in order to identify potential priorities for the engineering business featured in case study 4.

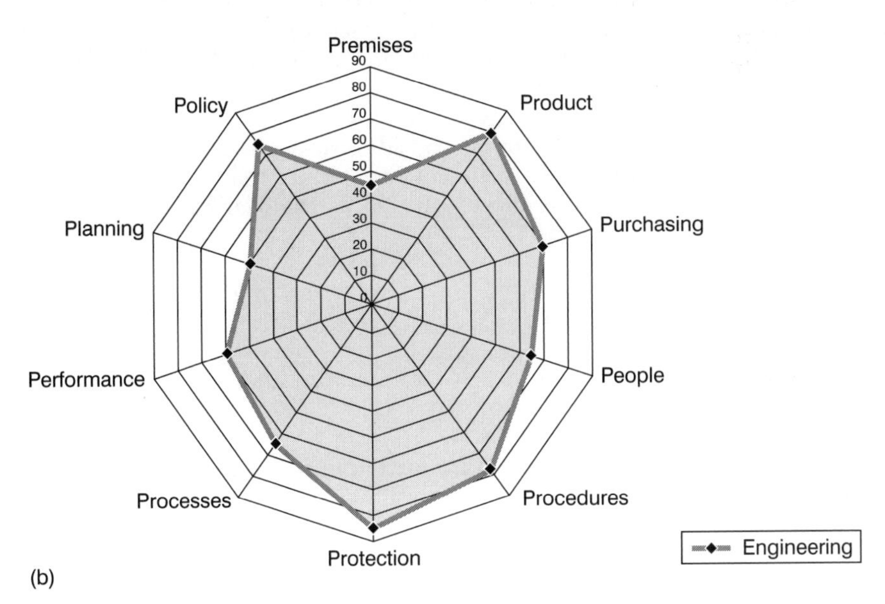

Figure 6.4 (b) Total scores

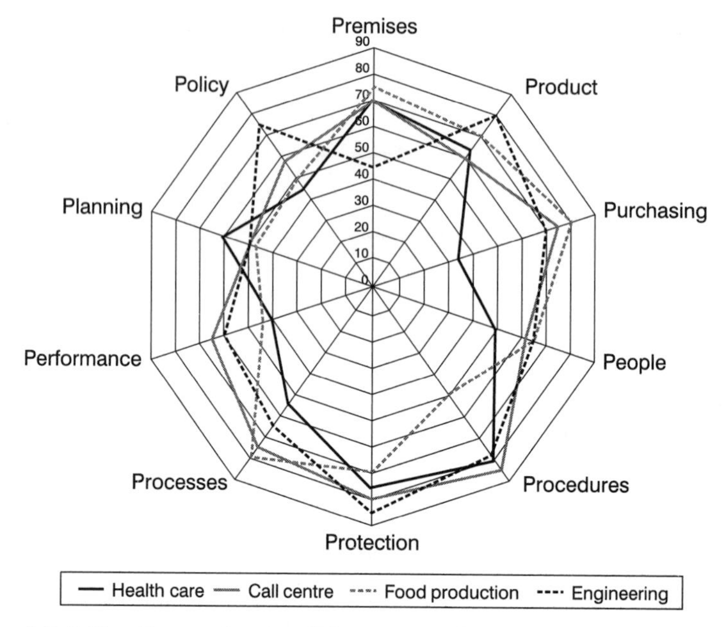

Figure 6.5 Spider diagram showing all four case study scores

6.5 Strategic considerations for case study firms

Performance can often become a negative risk factor as it potentially introduces additional criteria that impact on existing factors. On the other hand, it is an important management tool to identify and eliminate unnecessary or uneconomic elements of the business operations.

The next sections of the book consider strategic questions in more detail, but it is worth noting here some points that will need closer reflection by firms such as those featured as the four case studies.

Health service businesses

The immediate concerns relate to competitive and financial strategic issues, such as:

- a commitment to improving administrative and support structures in order to continue to provide relevant, accessible services to all client groups
- forward planning (five years and ten years) to decide the optimum target balance between private and NHS work given the existing inner-city location (no plans to relocate).

Call centre facilities

Employment and legislative strategic issues seem to be the main ones in this context, including:

- recognition and commitment at the most senior level to valuing the crucial role played by people in the organization
- combining policy/planning/performance elements in order to involve everyone in designing the process to improve job satisfaction, reduce potential risks from poor morale/high staff absence/high staff turnover
- involvement and consultation used to maintain loyalty and commitment of staff for optimum performance.

Food production and processing

There is considerable overlap between different strategic issues in this instance, as competitive, legislative and financial issues are of concern, such as:

- establishing and monitoring systems to ensure consistency, quality and traceability of supplies for the future, including alternative options in unforeseen or emergency situations (such as BSE/foot and mouth/flooding)
- a review of risks associated with location, particularly distribution channels and methods.

Engineering and manufacture

Competitive and legislative issues are of concern for manufacturing industries at a strategic level, particularly:

- the choice of a company-wide approach to quality/environment/health and safety issues irrespective of differences between sites, local conditions and legislative requirements, leading to consistency of approach/wider recognition by global customer base/plus standard provision of product or service
- commitment to use of sustainable, ethical supplier base for goods and materials.

Part 3

Chapter 7

Management strategies

7.1 Strategies for managing the risks

A comprehensive analysis of risks to the business is a vital step in confirming assumptions and gaining a full picture of the potential extent of harm possible if actions are not taken. However, this has really just been an auditing exercise up to this point and does not demonstrate that risks are being managed effectively or otherwise.

In Chapter 1, it was noted that the 10 Ps are intended to provide a range of prompts to ensure overall coverage of business activities without stressing the importance of one area over another. It is clear that, although a useful model or tool to work with, the boundaries are not so easily separated out from each other. Whichever element is used as a starting point for the earlier assessment stage, there will be knock-on effects on all other elements.

While some elements are considered to be 'operational' rather than 'strategic', they are just as important in the context of managing risks to the business. All elements of the three central groups of physical properties/people/process feed directly into the planning process and thus into policy-making decisions. So far, these two elements of planning and policy have only been discussed in relation to potential risk factors rather than as elements of the management process itself.

The prioritizing activities carried out at the end of Chapter 5 are, of course, part of the planning process and rely on any internal shortcomings in this element of the 10 Ps being corrected in order to ensure validity of the process. Patchy, inadequate sources of data and lack of management skills or expertise in-house, are likely to have gained a medium to high-risk factor rating, thereby requiring fairly urgent action. Policy-making risk factors, on the other hand, are more likely to have attracted a lower rating on the basis that changes to policy will be far-reaching and long term rather than needing urgent, immediate attention.

7.1.1 Planning

The priority rating for risk factors needing attention has formed the basis of identifying future actions. It does not necessarily represent a set of targets or objectives for the business overall. Depending on the size and structure of the organization, the range of activities required in order to reduce or eliminate the risks could well be substantial. Planning activities need to take into account other factors in order to be effective and reach the targets set. These include questions such as:

- actions required
- where
- by whom
- by when
- resources required.

Actions required

Actions required could range from:

- more in-depth, technical analysis of potential risk factors identified, such as noise levels or air contamination, possibly on a regular surveillance basis over a specified period of time
- replacement of existing machinery or equipment
- sourcing new or existing suppliers to find alternative, safer products or materials
- provision of training for specified groups of workers
- recruitment of internal staff, or making contact with external sources, in order to up-grade skills or knowledge base and expertise
- sale, purchase, or renovation of premises and plant
- relocation of some business activities to reduce potential risks from over or under capacity utilization
- evaluation of alternative sources of finance
- establishing more effective, wide-spread consultation mechanisms that include all relevant groups of workers.

Where

Where and which business units directly affected?

- current and proposed location of business activities, facilities, staff
- where to access technical expertise, either centrally provided or choice devolved to individual business units
- is in-house or external provision most appropriate in each of the risk areas identified?

By whom

Who will be involved in the activities required?

- allocate responsibility for ensuring actions are taken as planned
- devolve authority adequately to ensure decisions are made and actions taken
- arrange for results of activities to be monitored effectively, agreeing criteria for measurement and time scales for each stage

- make sure everyone is involved directly in discussions, informed of outcomes and decisions made and given relevant feedback on progress
- confirm the skills or expertise required to carry out the set tasks competently and that those with responsibility for actions possess such skills or expertise.

By when

When are activities scheduled to take place?

- agree urgency of actions required and specify short- or long-term time scales
- identify the milestones and target dates for monitoring/review/evaluation of results
- make appropriate budget forecast proposals and fit with investment plans, cash flow patterns etc.
- make sure these fit with legislative requirements.

Resources required

Vital to ensure adequate resources are available to enable successful outcomes:

- allocate sufficient funding to ensure the best fit with improvements required, available at the right time to ensure optimum results
- ensure additional time is available for additional roles and responsibilities, at all levels of staff
- establish procedures to provide access to data, remembering it needs to be timely, accurate, valid and relevant for the tasks set.

The plan(s) of action produced must be clear to those directly involved and to observers, 'transparency' being a key word in this context. Commitment to the planned actions is vital, right from the earliest stages and is most likely to be forthcoming when all those affected by the decisions feel that they have had an input to the decision-making process. This is true at all levels of the organization and transparency of management decisions is increasingly seen as a significant factor in the judgement of the success of an organization by its various stakeholder groups. It is not just in the context of health and safety that consultation with workers or their representatives is expected and enshrined in the regulations, or that people should expect to receive sufficient information about the firm's activities.

It is much easier to measure and evaluate outcomes of planned activities and to judge whether objectives and targets have been met if these are clearly set out at the earliest stages. Such targets or objectives must not be confused with the overall aims, which are likely to be much broader in content and context. For instance, the aim may be to improve the skill levels of staff in a particular division, but in order to judge whether this has been achieved, more specific objectives need to be stated, such as:

- by the end of six weeks a training needs analysis (TNA) will have been completed for all specified workers and gaps in individual skills or competences identified
- a relevant training plan will be prepared for each individual and learning targets set
- by the end of three months appropriate training providers will have been identified and a 12-month training programme agreed.

Although some of the risk factors may have been given a very low rating and considered to be comparatively trivial or low-level risks, a review of the situation must be carried out within a reasonable time frame – probably annually – in order to confirm that conditions have not materially altered. Other changes introduced must be evaluated to check that they have not had a negative impact on existing conditions and that existing controls are indeed adequate and sufficient.

In high staff-turnover environments, introduction of new workers to a job or department may increase the risk factor rating even if all other existing conditions remain the same. Factors considered earlier when deciding priority ratings against individual risks are just as important in the subsequent planning stage. To reiterate, these included:

- severity: the extent of harm, injury or damage likely as a result of exposure to the risk and of not taking action to reduce existing risks; minor or small-scale damage, localized impact, inconvenient loss, compared with severe, irreversible damage or injury, rapid and wide-ranging loss
- who will be affected: number and spread of individuals that will be affected, at departmental, business unit or company-wide level; impact on the wider community and environment
- immediacy: likelihood it will happen and possible time frame, inevitable in near future, or much longer before the impact materializes
- disruption: does the risk represent significant disruption to business activities, either in the short or long term, perhaps just a minor 'blip', or a fundamental change to the way the business is organized
- ability to recover: does the organization possess the necessary internal skills to be able to deal with the risk; what are the strengths and weakness that will directly impact on the ability to recover; will it be able to recover at all
- cost implications of getting it wrong: increasing reliance on litigation by customers, clients and others; the impact on public image, shareholders' views and on the owner him or herself; financial cost to the business.

By this stage, we have covered the fifth principle of the risk assessment process identified in Chapter 2 – 'monitor and re-evaluate after appropriate time scales and when circumstances/materials/processes etc. change'.

7.1.2 Range of strategic approaches for dealing with risks

It will be clear by now that the approach taken in this book is aimed at developing a holistic, interdepartmental system that tries to reduce the negative impact of functional boundaries and the lack of effective communication often associated with such barriers. In addition, the emphasis is on tackling all potential risks to the business equally in the first instance, to try to overcome problems associated with financial risks taking precedence over operational-level factors related to health and safety and other legislative requirements.

The approach taken by individual organizations will often stem from the attitudes, beliefs and to some extent the functional area of expertise of the founder or owner of the firm. They will have stamped their own ethical beliefs on the way the firm is organized, the priorities it gives to different risks facing the firm and the way internal structures are developed. Reference is often made to the 'health and safety culture' within a firm and the overt methods for protecting workers or others. At a broader level, the internal culture will also impact on the way responsibility, authority and blame is levelled at individuals or business units.

These philosophical beliefs underpinning the whole business are not always expressed verbally, apart from selected phrases perhaps in Annual Reports, or as Mission Statements that are intended to be more 'inspirational' than confirmatory. It is, therefore, worth reiterating senior level commitment to continuous improvement, effective management of risks, operating legally and without detriment to the individual, the wider community or the environment. Ideally, such commitment should be based on:

- an employment strategy that makes best use of the human resources available internally and externally
- a management strategy that reflects the organization's commitment to working within national and international legal parameters in all areas of operation
- a marketing strategy that provides a product or service that consistently meets all specifications set and is best suited to the customer's needs
- a financial strategy that balances an appropriate pricing strategy with investment decisions in order to optimize returns and ensure the continued life of the business.

In this instance, there is only room to mention these facets of the firm briefly, while acknowledging there may be fundamental problems associated with the need to change internal attitudes and beliefs in order satisfactorily to manage all the risks. It is hoped that if this is a major issue for an organization in relation to the way they assess and control risks, it will have appeared as such in the sections on employment and competitive risk factors and received a score that reflects this.

Other approaches to consider include:

- the use of industry standards and 'bench marking' to judge performance
- formal management system standards, such as the ISO or BSI ranges
- continuous improvement systems and schemes such as investors in people (IIP)
- financial strategies.

Use of industry standards

Some industries have identified 'good practice' over many years and produce guidance standards for use by sector players. They are generally voluntary, although may be mandatory in some instances, and produced by those working or experienced in the sector to ensure they reflect the conditions most likely to be found. For those where a licence to operate is required, such standards may form the basis of licence applications, to demonstrate awareness of and commitment to good practice.

As they are targeted at specific industries, they are often developed from the operational level, with less emphasis on strategic or management issues. In this case, some of the risk factors may be missed at the evaluation stage or assumed to be dealt with differently at senior management level. However, they still provide a valuable source of guidance for individual firms and a useful structure from which to develop the overall management system.

Bench marking, where a firm's performance is measured against other players in the industry to judge how effective they are, has grown over recent years. Although generally associated with larger organizations, there may be value in smaller or medium-sized firms checking their own progress against others in order to develop future strategy for growth. The process includes some form of auditing or reviewing existing conditions in order to identify gaps and set targets for improvement. Bench marking tools are often considered to be motivating for workers and other stakeholders, provided everyone is clear about what the intended outcome from the exercise is. The CBI, among others, has developed useful tools to carry out such evaluations, as has the HSE in relation to health and safety performance.

Use of formal management system standards (MSS)

Many larger organizations, certainly in manufacturing and some business support areas, use formally structured MSS that have been developed by standards bodies, with input from relevant industry representatives. Apart from standards related to specific types of activities, these MSS are intended to be generic and usable by a wide range of business enterprises. The most commonly used ones in the UK are for quality management (as the BS EN ISO 9000 series); environmental management (as the BS EN ISO 14000 series); and guidance for managing health and safety (as the British Standard BS 8800).

They provide a structure for evaluating the organization at various stages, for setting parameters within which the level of production/performance will be expected to fall and for establishing a system to monitor, measure and record performance at these specified stages. Each has until now taken a slightly different approach to structuring the system, as shown below:

- Quality MSS – developed primarily to demonstrate consistency and to achieve and maintain a specified standard of performance rather than as a basis for taking action to minimize broader risks to the business, such as H&S. They are aimed at achieving customer satisfaction with the product or service and managing risks associated with getting it wrong.
- Environmental MSS – primarily intended to help firms develop objectives and policies to ensure environmental aspects of the business are managed effectively, taking into account the relevant regulatory requirements. After an initial 'status review', it follows the plan/do/check/act model of policy – planning – implementation – checking and corrective action – management review, leading through the loop to continual improvement.
- OH&S standards – BS 8800 is a guidance document rather than a formal MSS, but is structured to fit more closely with the BS EN ISO 1400 outline as well as the HSE approach suggested in HS(G)65 publication. This follows the review – policy – organizing – planning and implementing – measure performance model, although this is more difficult to fit into existing systems based on the BS EN ISO MSS. The OHSAS 18001:1999 specification produced by BSI follows the BS 8800 pattern quite closely.

Each of these systems has advantages and leads the organization into developing management strategies that aim to identify and control risks, although at the time of writing, they are each stand-alone models. There has been pressure for some time for all three elements of quality, environment and H&S to be integrated into a single MSS, although this has been resisted from the small firms' perspective on the basis that it may then become too cumbersome or bureaucratic to operate. Of greater concern is the polarization of views in relation to the need for external certification for such systems and the inevitable additional costs to the business that this entails.

In reality, it may be that an integrated approach to managing all the diverse risks to the business is more appropriate than an 'integrated management system standard'.

Continuous improvement and schemes such as IIP

There are many models and schemes to help firms develop their internal management systems, often focusing on specific elements of the business rather than the full range of factors. It is likely that even the best-run organizations will have some elements of the business that could be improved, so the notion of aiming for ever better performance rather than becoming complacent once a specified level has been reached seems an attractive one.

Formal schemes such as investors in people (IIP) help to focus attention on the value of people to the organization and to encourage positive action at all levels of the firm. However, they too are often limited to certain areas of business operations or structures, require significant resources to support and ignore their impact on other areas. IIP, for instance, confirms the value of staff to the successful operation of the firm, but does not include reference to health and safety or protection from these risks.

Financial strategies

The cost implications of taking a proactive approach to identifying and managing all the risks identified so far may be considerable. This is often stated as the reason why firms are reluctant to start on the path of evaluation of risks, preferring to ignore them and hope nothing happens to force them to take action. Clearly, there are likely to be costs associated with some of the actions required to reduce or eliminate risks, but these must also be considered alongside the potentially higher costs of trying to deal with the damage that could result from non-action. There will, therefore, need to be some form of cost/benefit analysis carried out to support and justify decisions made.

The use of formal management systems such as those outlined earlier does represent a cost, certainly a net cost in relation to environmental management in some cases and this may relate to the type of industry sector rather than size of firm. More problematic is the need to accommodate different levels of risk across international boundaries alongside different levels of control expected in each location. The most successful multinational organizations have found it easier to operate with the same expectations across boundaries, given the increases in public pressure for firms to take an ethical, sustainable approach to operating their business.

Financial strategies may need to include reference to:

- divestment or acquisition of companies, business units or plant
- developing in-house capabilities or outsourcing stages of production
- investment programmes and time scales
- use of hire purchase or leasing rather than outright purchase
- use of credit reference agencies and status reports for clients/ suppliers
- use of factoring and discounting for bad debt.

The need to use such strategies should be clearer if the whole range of risks has been considered and impact assessments made for proposed actions. As noted earlier, for very large organizations, these choices will be made from a base of significant experience in financial management, but for smaller operations this expertise will not be so readily available.

In the next section, controlling risks and the importance of different stakeholders will be discussed further, with policy development in section 7.3.

7.2 Stakeholders and spreading the risks

We have talked about controlling risks in a practical sense and have spent a lot of time coming to grips with the full range of risks facing the business. At the strategic level, the issue is more complex than just eliminating or reducing the risk as much as possible in order to control potential loss. Measures to avoid risk involve the range of controls identified earlier, so controlling the risk at source, for example by reducing the likelihood that a fire will occur, and measures to reduce the risk or resultant loss include things such as fire-fighting equipment and procedures in case a fire actually starts.

There is also the question of how to finance the potential losses from remaining risks, once everything possible has been done. Small-scale events and losses are likely to be frequently experienced by firms, but the scale of real costs is not fully realized or is overlooked. On the other hand, experience of a serious event is rare in most firms, so the impact is often underestimated. Clearly, if the risk relates to personal injury or death it has to be reduced to as low a level as possible, but for other types of risk factors a decision may have to be made to accept the potential loss rather than utilize more resources in order to reduce it further.

The cost/benefit analysis referred to earlier is often problematic in relation to health, safety or environmental risks, largely due to the difficulties in quantifying the potential loss and benefits in financial terms and the time lag between expenditure and resulting reduction in loss. To some extent, this will be determined by the type of enterprise, industry sector, level of contact with the end-user of the product or service and the various groups of people who have an interest in the firm – the stakeholders. For our purposes here stakeholders include:

- workers and employees who have an interest in being protected from risks in the workplace and in the retention of jobs
- worker representatives at all levels, whether part of recognized unions or not, who have an interest in maintaining levels of protection for workers as well as the firm's continued existence
- senior management and/or board of directors, who hold ultimate responsibility for the resultant losses from mismanagement of risks and may hold personal liability for breaches of health and safety law as well as corporate liability
- the owner, whether as an individual or holding company, who will want to see the optimum use of all resources and potential risks managed effectively in order to reduce losses
- shareholders who have invested resources in some form into the business and expect potential losses to be as low as possible in order to gain the expected return from their investment

- banks and other financial institutions who need to be sure that borrowing is correctly geared relative to turnover, profit and investments and that repayment is secured in some way
- enforcement authorities, both local and national, who expect to see evidence that regulatory requirements are adhered to, proper records kept and some form of management system exists to safeguard individuals, the public and the organization from risks
- insurance providers who expect that all risk factors have been identified, adequate controls are in place to eliminate or reduce potential losses and insurance cover provided is appropriate and sufficient
- the public, whether as consumers, part of the local community, or as the community at large, concerned that risks arising from the activities of the organization are adequately controlled in order to protect them from wider losses.

Perception of risk is an interesting factor in relation to different stakeholder groups, influencing their views on what is considered an 'acceptable' level of risk. Some older, traditional industries display a level of acceptance of safety or health risks that is considered to be totally unacceptable by others. The issue of whether the risk is taken voluntarily – that is, as a matter of choice by the individual – is also a factor, particularly in the grey area between work and life-style choices such as smoking or sports activities. An acceptable balance is very difficult to achieve in the leisure industry, for instance, where customers expect to be able to smoke if they wish, but workers need to be protected from the perceived risks associated with passive smoking.

The question of control over the situation or choice by the individual is, therefore, a crucial factor of perception and acceptance of risk. An individual trading on the stock exchange via the internet may expect and accept considerably higher losses from investments than if funds had been placed with a recognized investment body.

Losses are often divided into two main groups of consequential loss and direct loss.

1. Consequential losses are more difficult to quantify, are not always immediately apparent and therefore tend to be overlooked or underestimated. They include lost time and business interruption caused by accident investigations, machinery breakdowns, transport delays, supplier problems, disputes with workers and external events, such as flooding. They may also include other elements such as cash flow difficulties and late payment of debts, loss of records and important data, as well as losses associated with bad publicity and poor public image following a serious event and fines or penalties.
2. Direct losses are generally the more obvious ones related to replacing equipment or machinery, repairs to plant and premises, damage or loss of goods, payment of third party claims.

Broadly speaking, the risk of direct losses is transferred to an insurer but consequential losses are uninsurable and retained by the organization. If small-scale losses occur frequently, they are probably retained by the company as increased premiums based on claims record may be disproportionately higher than the value of losses incurred. In addition, some non-frequent severe losses may also be uninsurable, such as flood damage or bomb explosions.

Speculative risks, where the outcome may be a win or lose situation, are unlikely to be insurable as are losses due to inadequate security measures being in place, such as theft of cash from open tills rather than while stored in a locked safe. Residual risks can be spread through a combination of transfer (to an insurer) and retention (within the organization), perhaps by covering the first part of a claim by an excess payment, or by part-insurance where a percentage of the risk is covered.

In reality, many firms rely too heavily on the insurance route to protect them from losses, mistakenly believing that all the risks are covered including consequential loss, rather than taking appropriate action to avoid or reduce the risk in the first instance. It is important, therefore, that firms appreciate the different ways they can spread the impact of potential losses and make a realistic appraisal of the real losses likely to occur from residual risks identified against each group of risk factors. While some losses associated with legislation or security risk factors are likely to be insurable, this is less likely to be the case in relation to competitive and financial risks or indeed some aspects of employment.

7.3 Policies

Legislation requires that a policy on health and safety must exist and be recorded if more than five people are employed. A typical list of features in such a policy for a smaller firm might include commitment to:

- providing a safe and healthy work environment for people
- ensuring premises are maintained properly, good housekeeping standards are kept and adequate facilities are provided for workers and others on-site
- producing a product that does not jeopardize the safety and health of others, or the environment
- purchasing less hazardous raw materials where possible, that are healthier, safer and more environmentally friendly to use
- identifying hazards and assessing risks to workers and others who may be affected by activities of the firm
- involving workers directly in discussions about health and safety issues or concerns, to ensure their input and commitment to working together to tackle these issues
- providing sufficient resources, information and training to people to ensure they can carry out their duties and fulfil their responsibilities in a healthy and safe manner
- establishing procedures for work in all activity areas of the business that take into account the health and safety protection of workers

- making sure that procedures intended to safeguard people and the environment are followed correctly and that people are adequately supervised
- ensuring suitable monitoring and recording systems are in place
- providing adequate protection for people against the risks of damage to health and harm or injury, resulting from work activities or fire
- ensuring processes are carried out using equipment and machinery that is appropriate, as safe to use as possible and properly maintained
- setting targets to reduce where possible accidents and ill health in the workplace
- regularly reviewing the situation to see whether targets have been met, existing controls are still adequate and in place, or new targets need to be set.

This list is fairly compact but is probably sufficient to identify intent on the part of the firm to manage their health and safety responsibilities in a positive way. It could be elaborated to cover other areas such as smoking, use of drugs, bullying, fire protection and would need additional references to details about individual responsibilities and internal procedures. The list broadly follows the 10 Ps elements, although it does, of course, concentrate exclusively on the health and safety risks.

However, for more complex organizations and in the context of risk management generally, it is clearly insufficient. The risk factors have been identified and evaluated in detail, risk control measures considered and required remedial actions prioritized and management strategies developed. It is vital that appropriate policies are established to support the agreed strategies, that everyone knows and understands them and that their effectiveness is monitored over time. As we saw in Figure 1.2, the elements referred to as 10 Ps all interact with each other, feeding into the planning and decision-making process and the development of relevant policies. It is a continuous process, requiring regular monitoring/review/re-evaluation to ensure it reflects the dynamic nature of the business, rather than being a one-off activity of limited value to any stakeholder groups.

The following section highlights some of the areas where a policy statement is needed in order to confirm commitment at the most senior level within the organization.

7.3.1 Premises

The policy issues related to premises are likely to relate to facilities provided and geographic location, such as:

- provide a healthy and safe working environment, with adequate facilities for workers
- maintain buildings, site and surrounding area as necessary
- use local sources of labour where possible, providing adequate relocation packages to assist workers and their families who need to relocate

- ensure adequate site access for people with physical disabilities, whether workers or customers, taking into account the existing physical structure of buildings
- introduce a smoking policy to accommodate the needs of smokers and non-smokers
- ensure all legislative requirements and licensing conditions are fulfilled
- safeguard the site and surrounding areas from potential harm or damage from fire, pollution, noise and other recognized hazards.

7.3.2 Product or service

A wider range of issues needs to be addressed under this heading, given the broader concerns of the end-user as well as the organization as provider of the service, or manufacturer of the goods. These will include:

- comply with requirements of working time regulations to ensure adequate breaks etc. for staff
- provide a safe product or service for customers, with relevant information and adequate packaging
- provide environmentally-friendly product or service as far as possible, taking full consideration of safe disposal of obsolete products or waste
- comply with relevant consumer protection legislation and offer advice/information/support through customer support service
- ensure proper training and supervision is provided to workers to enable them to produce a product or service that conforms to specification
- safeguard access to sensitive data and valuables, whether on-site or in transit
- operate according to the Late Payment of Commercial Debts (Interest) Act 1998, ensuring payment terms are clear to all parties.

7.3.3 Purchasing

It is important to identify policies related to identifying, accessing and paying for supplies as well as the logistical elements of delivery and storage. Policy statements are therefore likely to refer to:

- ensure prompt payment of bills to ensure consistency of supply
- avoid use of supplier conditions that restrict the supplier base unnecessarily (The Competition Act 1998)
- establish system to check quality specifications for supplies are met, appropriate to type of goods or services being supplied
- ethical purchasing policy based on use of suppliers that operate within the law and do not endanger the health or safety of their workers or the environment

- compliance with conditions of existing internal management system or other standards
- ensure safe storage, handling and transport of goods
- use the most cost-efficient purchasing and delivery methods
- seek to find more sustainable sources of raw materials and substitute with less hazardous substances where possible.

7.3.4 People

Given the considerable raft of legislation in place related to worker protection, it is vital to reflect these responsibilities fully in the policy of the organization as a means of supporting the organization's aims and goals, not just because the law requires it. While the detail for such policy statements can be developed from guidelines produced by, for instance, the Equal Opportunities Commission, the following gives a picture of the broad sweep of issues that need to be included:

- make a positive commitment to equality of opportunity in recruitment/training/promotion practices
- strive to reflect the local population mix within the workforce and avoid discrimination on the grounds of gender, race, religion, age etc.
- make adequate provision for the employment of people with disabilities where necessary and develop a rehabilitation programme to enable return to work as quickly as possible
- provide access to relevant training, information and guidance to ensure people have the necessary skills and competence to carry out their job satisfactorily
- follow family-friendly working practices where possible, to offer flexibility and encourage worker loyalty
- intolerance of workplace bullying or threatening behaviour, aiming to reduce potential for unnecessary levels of stress
- comply with the requirements of employment and worker protection legislation in spirit as well as to the letter.

7.3.5 Procedures

Policy statements relative to procedures within the organization will need to reflect the individual nature of the business, but are likely to include the following:

- consult with workers and others on issues that affect them in the workplace, providing opportunities for them to elect representatives as necessary
- establish structures for worker representation that recognizes the role of union representatives where these exist
- ensure discipline and grievance procedures are developed with the intention of protecting people and ensuring fair treatment

- provide health surveillance where necessary and access to relevant occupational health services that exist locally, nationally, or internal to the firm
- ensure safe systems of work and all procedures designed to safeguard health, safety, security of workers are always adhered to
- comply with requirements of any management system standards that are in place
- ensure that reporting procedures are correctly followed, whether internally required or for external purposes such as reporting under RIDDOR
- comply with the requirements of relevant legislation, whether Employment, Health & Safety, Environment, or more general regulations such as Competition Act 1998.

7.3.6 Protection

Policies must, of course, reflect the need to protect individuals, communities, property, assets and the wider environment, so may well include quite diverse elements such as:

- assess health and safety risks to workers, customers, others to ensure adequate controls are in place to reduce risks to as low a level as possible
- provide necessary personal protective equipment where other control measures cannot reduce risks any further
- take sufficient security measures to ensure protection of assets
- ensure funds are invested prudently and ethically to safeguard the future of the organization
- protect data in whatever format, sensitive information on workers or clients, according to the requirements of the Data Protection Act
- assess risks and ensure adequate controls are in place to protect the local and wider environment from damage due to the firm's operations – for instance emissions, fire, pollution
- make efficient use of energy sources to reduce the impact on climate change.

7.3.7 Process

Many of the policy statements developed already will cover aspects of the processes used, such as 'product' or 'procedures', but the following are also policy positions that need to be specified:

- ensure plant and equipment is appropriate for the job and regularly maintained to ensure safe and healthy operation
- replace and up-grade as necessary with safer, cleaner versions of plant to maintain optimum levels of production
- develop a waste management policy to reduce levels of waste/scrap materials produced and to dispose of them safely

- substitute less hazardous substances and materials used in production or provision of service
- identify storage/transport/distribution economies where possible.

7.3.8 Performance

In order to ensure transparency and commitment company-wide, it is worth stating the policy related to setting targets, identifying performance criteria and measuring outcomes in the following areas:

- agree and define criteria for individual/team work targets and performance measures that are relevant, achievable and measurable
- establish targets to reduce levels of waste/scrap
- establish financial targets and performance measures for future operations and growth
- ensure consistency of measures and data collection procedures across all business units
- monitor returns on investments regularly to ensure optimum returns are made within targets set
- comply fully with corporate governance requirements, based on transparency, agreement, commitment at the earliest stages
- review performance against specified criteria in all areas of business operation on a regular basis, on a timely basis to ensure appropriate remedial action can be taken swiftly to limit potential damage
- to maintain a positive image and standing within the local community where the firm operates.

7.3.9 Planning

As we have already seen, the planning process combines feedback from several quarters and policy statements should identify how this is to be managed effectively, for instance:

- incorporate input from all sectors of the organization where issues will impact on their working situation
- ensure adequate contingency plans to safeguard workers, business operations and the wider community
- make sure everyone knows what these contingency plans are
- establish and maintain workable communication channels throughout the organization
- take necessary actions to minimize the potential for fines, penalties or prosecutions against the organization or any of its component business units.

It is quite likely that there will be additional policy issues included to reflect the unique situation in individual organizations, but the list above covers the main policy areas that should be addressed. In addition, health and safety policy statements need to identify specific details such as names of people responsible for investigating incidents, first aid personnel, who to contact in an emergency, who are deemed to be 'competent persons'. Whether business unit specific or overarching company-wide policy statements, these health and safety policies must also include the name of the most senior director with personal responsibility for ensuring compliance with relevant legislation and that health and safety issues are given due priority in management decisions.

Once these policy areas have been considered and statements drawn up to establish them within the organization, they must of course be brought to everyone's attention. Ideally, people at all levels of the firm will have been involved in discussions about their content and have therefore demonstrated some commitment to the principles enshrined within them. This commitment must be demonstrated from the most senior levels down, as well as the shop-floor up, in order to maintain the momentum and ensure cynicism does not creep in.

In the context of this publication, a critical element of the exercise is to ensure the whole range of risk factors is considered equally, without greater emphasis on one aspect over others such as financial, health and safety, or human resource risks, depending on the interest area of those carrying out the assessment. It will be clear by now that the greatest challenge for most businesses is to combine the full range of strategic and operational concerns in such a way that protection against existing and potential future risks is an integral part of the management process.

Chapter 8

Conclusions

There is no single right way to manage the myriad of risks facing business today, given the considerable changes in working patterns, globalization and the unprecedented increase in the use of electronic media to support operations during the last decade. The 10 Ps approach aims to bring together the most common elements in a way that recognizes the importance of all risk factors to the successful operation of the firm and cuts across management functional boundaries.

Clearly, the risk management skills embodied within these different functional areas have a crucial role to play in the stages of the process described in Chapters 2–6 and must all be brought together to make best use of them. Whether public sector agency, large or small private-sector enterprise, or charitable trust, the risk factors identified are likely to apply to a greater or lesser extent to all of them. In particular, the vast majority of health and safety, environment and fire legislation applies equally to all but the smallest of enterprises, so risk assessment in these contexts is a valuable tool to demonstrate just how far-reaching the potential harm to the whole organization can be if existing controls are ineffective.

8.1 Identifying the risk factors

On reflection, Part 2 of the book is a comprehensive section that considers the risk assessment process in some detail, both for those new to the process and those with experience of assessing risks in just one area. Many of the principles are the same, of course, but hopefully additional points have been raised that may have otherwise been overlooked. It is always useful to review existing risk assessment procedures periodically in order to ensure they still fulfil their original intentions adequately and to reassure people of continuing senior-level commitment.

Experience has shown that a site plan is a good base from which to start the hazard identification process, helping those who are already working on-site and those who are bringing a 'fresh eye' to the risk assessment process in order to ensure risk factors are not overlooked due to familiarity. It also reinforces the whole-site, inclusive nature of the assessment and factors other than physical risk factors may become more apparent. For example, practical constraints on providing segregated smoking areas will generally be site-specific and joint use or ownership of premises is often accompanied by conflict over responsibilities for losses arising from uncontrolled risks.

Legislative risk factors impact widely relative to premises, as they do in relation to workers and others on-site, so this is often the initial focus for firms wishing to spread potential loss through insurance cover. As we note in Chapter 7, on its own this does not necessarily lead to sufficient protection.

At the time of writing, the second stage of the Disability Discrimination Act related to access to premises is due to come on-stream, the Climate Levy charge will soon be applied to UK businesses according to their power consumption levels and existing asbestos regulations (with their building management requirement to identify potential locations of asbestos materials in building structures) are under review. Potential losses for firms mismanaging these elements alone could be extremely damaging.

It is reasonable to assume that the choice of product or service provided by the organization is much more within their direct sphere of control, as are the risk factors associated with this choice. However, as this is so closely tied in with purchasing decisions, processes used and protection of people whether workers or customers, the risk factors are extremely far-reaching. Demographic changes impact on the range and type of goods or services being sought out by customers, but also impacts significantly on the make-up of the workforce in some industries. The opening by a major multinational firm of a day-care centre for older relatives of staff, rather than a crèche, reflects the potential problems facing many workers now faced with these additional responsibilities and is an excellent example of a positive risk management strategy.

Greater willingness on the part of consumers to seek compensation when unhappy with the goods, a rapidly expanding base of legal firms willing to help them and a substantial base of consumer protection legislation on content/structure/packaging of the product, all combine to represent a powerful risk factor for most industry sectors. While these consumers may be better informed than previously, media coverage is not always as positive or helpful as some would like and often fuels concerns about misinformation.

Although there is still debate about the causes of climatic change at a global level, pressure will continue to mount for industry to demonstrate that it takes the issue seriously and is aware of environmental concerns that people have. Ethical purchasing is an issue worthy of consideration, though not necessarily clear-cut in its application, as is the need to ensure fair and equitable trading conditions that stay within the guidelines of the Competition Act 1998. At the smaller end of the enterprise scale this may not represent as great a risk factor as it is likely to at the large-scale multisite end of the business spectrum.

Changing work patterns and increased worker protection legislation over recent years both represent potential risk factors for firms, not least in the need for ever more complex systems in place to ensure adequate protection for individuals. Flexible working and non-standard employment contracts make it difficult to track workers, to involve them in meaningful discussions about issues that effect them when they are mainly off-site and to ensure they receive sufficient skills or health and safety training. Flexibility may have been the by-word for competitiveness during the 1990s, but the downside is that it also equates with uncertainty that can in itself lead to greater stress.

There is increasing evidence of workers exposed to violence at work, particularly when working in face-to-face situations with the public and safety while driving for work purposes is taking on greater significance. All of these issues represent risk factors for the organization as well as the individual, with the attendant potential for loss that must be managed properly and fairly. As we have seen in Chapter 2, a crystal ball would be extremely valuable in relation to assessing the potential risks associated with use of substances or processes that may, in the future, be deemed to be more harmful than previously thought, leading to possible claims for damage after a 20- or 30-year time gap. Having said that, it is likely that everyone can think of examples where workers have complained about unacceptable side-effects or ill health while carrying out certain tasks, but legally they are considered 'safe' at present.

Finally, the risk factors identified at this stage of the process rely on the collection of valid, relevant data. Historical data such as absence records, accident reports, previous years' figures and returns, all have their place, but it is hoped that the wide range of suggested factors included here has led to the use of a wider variety of performance measures to ensure the comprehensive coverage required.

8.2 Evaluating the risks

The use of the ten headings, the 10 Ps, is a useful tool for ensuring that the full extent of business operations are considered equally, particularly when combined with the human resource, legislative, security, competitive and financial risk factors. For those working in some of these areas on a daily basis, the risk factors may be glaringly obvious, but as it is likely other people will be taking part in a comprehensive analysis of the whole firm, it is important to state them clearly.

It is worth spending time, therefore, making clear what the potential harm, injury or damage is likely to be from the risk factors identified and who is most likely to be affected – whether as individuals or stakeholders or business units. Potentially, there may be quite tangible losses from some risk factors, particularly those where specific individuals are most likely to be exposed to the hazard. On the other hand, there are also likely to be less easily defined losses, such as public image or credibility in the local community, as well as far-reaching cost implications for access to markets or services in the future.

How immediate the impact or loss will be should also play a part in the equation, given that the impact may be immediate and fairly low-key or indeed be a long-term problem that could lead to a fundamental change in the way the firm operates. In practice, the remaining factor is an important one, in that the likelihood that the potential harm will materialize, devastating as it might be, is actually very small.

The risk rating approaches suggested in Chapter 4 and Chapter 5 are typical for this type of activity, requiring a subjective evaluation to some extent. While computerized assessment systems can be used just as successfully, in their absence the approach suggested will provide a comprehensive, consistent picture to enable decisions to be made about prioritizing future actions. A numerical scoring system gives an opportunity for finer gradations between ratings, but may become cumbersome if a substantial base of risk factors is being addressed or the scoring range is too great.

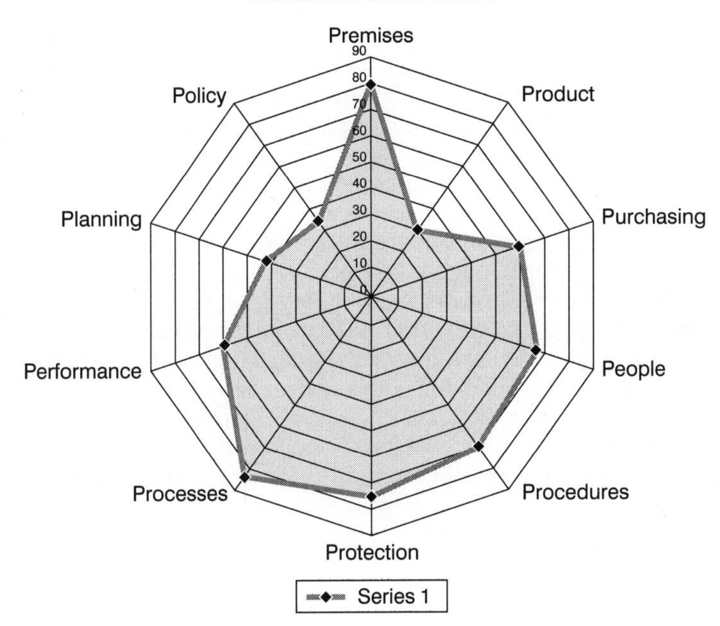

Figure 8.1 Total score for risk factors

In Figure 8.1, the total scores for risk factors identified in each of the 10 categories have been plotted in a spider-chart format (suggested by S.Fulwell, 2000) to illustrate exactly where the biggest risks are concentrated. As you can see in this example, the most significant risks relate to the processes used and the premises themselves, with much less concern evident in relation to the product itself and the management structures in place. Clearly, there needs to be some fundamental revision of the operational side of the organization, with some fairly urgent attention to risks posed by unsuitable premises or location.

8.3 Controlling the risks

In this context, we have considered physical, behavioural and organizational controls as a means of reducing the negative impact of exposure to hazards or risk factors, ranging from decontamination facilities to incentive schemes, recruitment policies to the use of formal management system standards.

Individual firms will have a diverse range of control measures in place, some more effective than others. Assumptions about existing controls should be questioned in order to confirm that:

(a) they still exist as they were originally intended to and
(b) they are still appropriate for the changing face of risk factors that challenge firms today.

Procedural controls in particular should be examined, as the fact that they appear in written form as part of contracts or procedure manuals does not necessarily reflect the extent to which they are followed. Insurance cover is an interesting point to consider in view of the risks identified, the extent to which existing cover encompasses these risks and the length of time since premiums and cover were reviewed and reassessed. The issue of insurable rather than uninsurable losses is a crucial one for discussion at this and later stages of the process.

As one of the primary reasons for carrying out risk assessment is to help in deciding priorities for future action, the work undertaken earlier must feed directly into this decision-making process. The most significant risks should emerge as those that require fairly speedy remedial action and others with a lower rating as maybe less urgent. However, as noted earlier even the lowest scoring factors still represent a need for action and lots of small-scale low-cost changes could represent a more positive impact on other risk factors, such as low motivation or morale of workers fed up with poor local working conditions they have complained about for months.

8.4 Managing the risks

The whole purpose of this book is to lead the reader through the risk assessment process, to a review of controls and identification of gaps in protection against loss or damage. That in itself may be an admirable stage to reach for some organizations where this has not been formally tackled to any great extent, except perhaps in the area of health and safety or competitive risks. The checklist shown in Figure 8.2 might be useful in those circumstances to confirm that actions have been taken and to demonstrate to interested parties that a comprehensive management approach is in place.

However, this is only part of the story and in effect demonstrates mismanagement of risks if no further action is taken to eliminate, reduce or spread the risks and safeguard the interests of all stakeholders in the business.

It is management's responsibility, whatever the size of firm, to take action to minimize the impact of losses on the business and to be seen to be protecting the interests of everyone, rather than one set of interests over another. Effective 'corporate governance' is a legal requirement and

Management actions taken:	Yes (tick)	In part	Complete by	Review date	Review by
a) Established priorities: – noted 'Review' where controls adequate – identified high-risk factors with Priority Rating 4–5 – identified medium-risk factors with Priority Rating 3 – identified low-risk factors with Priority Rating 1–2					
b) Prepared Plan of Action, with steps needed and time scales set for completion, for: – Priority Rating 4–5 – Priority Rating 3 – Priority Rating 1–2					
c) Established appropriate data bases and recording systems					
d) Provided staff with relevant and sufficient information to be able to complete tasks					
e) Established appropriate consultation procedures with workers					
f) Allocated and confirmed parameters of individual responsibility and authority					
g) Identified one or more 'competent person(s)' for Health and Safety risks					
h) Arranged methods for keeping up to date with legislative changes and their impact on the firm					
i) Prepared a full range of Policy statements, including Health and Safety					
j) Notified everyone of details of these policy statements, and confirmed commitment at the most senior level of the firm					

Figure 8.2 Management checklist

although reference is often made to directors in this context, it is not just applicable to incorporated enterprises (where the term director is legally recognized) but also to those at the most senior board level of other enterprises.

The main principles of the 'Combined Code of the Committee on Corporate Governance' include reference to the need to:

- maintain a sound system of internal control to safeguard shareholders' investment and the company's assets
- conduct a review of internal controls and their effectiveness, including financial, operational, compliance controls and risk management
- review whether they need an internal audit function if they do not already have one
- report findings to shareholders, including explanations of how the principles of this Code have been applied.

The Institute of Chartered Accountants in their Guidance for Directors[3] notes that this risk-based approach 'should be incorporated by the company within its normal management and governance processes', rather than as an exercise just to comply with the law.

In addition, the revised HSE Code of Practice on Health and Safety Responsibilities of Directors[4] notes that:

'Effective management of health and safety risks:

- maximizes the well-being and productivity of all people working for an organization
- stops people getting injured, ill or killed by work activities
- improves the organization's reputation in the eyes of customers, competitors, suppliers, other stakeholders and the wider community
- avoids damaging effects on turnover and profitability
- encourages better relationships with contractors and more effective contracted activities
- minimizes the likelihood of prosecution and consequent penalties.'

Clearly, these are all powerful arguments, irrespective of the legal imperative to comply and all businesses must take on board the need to assess, control and manage all risks to the business comprehensively, but also to do so in a transparent way that demonstrates their full commitment to the process. This in itself represents a further risk factor that must be evaluated and dealt with according to its priority rating.

The 10 Ps approach encompasses all the necessary elements to demonstrate full compliance with the needs of the Combined Code, involving an holistic evaluation of risk factors and taking positive management action to protect the interests of all parties against potential loss. In addition, it is intended to act as a practical trigger for those charged with responsibility for managing the risks without drawing too deeply on the theoretical background that underpins it.

Chapter 9

Useful references

1 Combined Code of the ICA Turnbull Committee on Corporate Governance, known as the *Turnbull Report*, in place from December 2000. ICA ISBN 1 841520101.
2 Jeynes, J. (2000) *Practical Health and Safety Management for Small Businesses*. Butterworth-Heinemann.
3 Institute of Chartered Accountants (1999) ***Internal Control*** – *Guidance for Directors on the Combined Code*. ICA ISBN 1 841520101.
4 *Health and Safety Responsibilities of Directors* (revised 2001) HSC consultative document on proposed Code of Practice CD167 C40 1/01.

Government departments

Health and Safety Executive

HSE Information Centre
Broad Lane
Sheffield S3 7HQ
Tel: 0114 289 2345 Fax: 0114 289 2333

HSE Books
PO Box 1999
Sudbury
Suffolk CO10 6FS
Tel: 01787 881165 Fax: 01787 313995

HSE Infoline Tel: 08701 545500
HSE Local offices around the UK: find their number in your local Telephone Book
On-line access:www.hsedirect.com
www.hsebooks.co.uk for on-line ordering

Department for Environment, Transport and the Regions (DETR)

Great Minster House
76 Marsham Street
London SW1P 4BR
Website: www.detr.gov.uk/hsw/index.htm
Environment Agency: general enquiry line Tel: 0645 333 111

- Pollution – prevention pays
- Producer responsibility obligations 1997
- Money for nothing – your waste tips for free
- Special waste Regulations 1996 – how they affect you
- A new waste management licensing system – what it means, how it affects you

Department for Education and Employment (DfEE)

For information on employing people with disabilities:

DDA Helpline Tel: 0345 622 633 or 0345 622 644

Department for Trade and Industry (DTI)

DTI Consumer Safety Unit
CA3a, 4th Floor
1 Victoria Street
London SW1H 0ET
Tel: 0207 215 0359 Fax: 0207 215 0357

For information on publications, contact:
DTI Publications Orderline Tel: 0870 1502 500
Useful references:

- *A Guide to Working Time Regulations*
- A detailed guide to the National Minimum Wage
- The Late Payment of Commercial Debts (Interest) Act 1998: A users guide
- Guide to the Consumer Protection Act 1987 – Product liability & safety provisions
- Variety of publications from the Patent Office, DTI on intellectual property protection

Department of Health

Department of Health
Wellington House
133–155 Waterloo Road
London SE1 8UG
Tel: 0207 972 2000

Department of Social Security (DSS) regarding sick pay or maternity pay

DSS Advice Service Tel: 0345 143 143

Office of Fair Trading

Field House, Room 107
15–25 Bream's Buildings
London EC4A 1PR

- www.oft.gov.uk
- enquiries.competitionact@oft.gov.uk
- What your business needs to know (about the competition Act)
- How your business can achieve compliance

Other useful sources of information

Association of British Factors and Discounters
1 Northumberland Avenue
Trafalgar Square
London WC2N 5BW
Tel: 0207 930 9112 Fax: 0207 839 2858

Association of Invoice Factors Ltd
20/22 Bedford Row
London WC1R 4JS
Tel: 0141 248 5100

or: Association of Invoice Factors Ltd
Finlay House
10–14 West Nile Street
Glasgow GT1 2PP
Tel: 0141 248 4901

The Chartered Institute of Arbiters
24 Angel Gate
City Road
London EC1V 2RS
Tel: 0207 837 4483

Institute of Credit Management
The Water Mill
Station Road
South Luffenham
Oakham
Leics LE15 8NB
Tel: 01780 721888

Finance & Leasing Association
18 Upper Grosvenor Street
London W1X 9PB
Tel: 0207 491 2783 Fax: 0207 629 0396

Chartered Institute of Purchasing & Supply
Easton House
Church Street
Easton on the Hill
Nr Stamford
Lincs PE9 3NZ
Tel: 01780 56777

Institute of Chartered Accountants
Chartered Accountants Hall
PO Box 433
Moorgate Place
London EC2P 2BJ

www.icaew.co.uk/internalcontrol
www.accountancybooks.co.uk

Institution of Occupational Safety & Health (IOSH)
The Grange
Highfield Drive
Wigston
Leics LE18 1NN

BMA
BMA House
Tavistock Square
London WC1H 9JP

Faculty of Occupational Medicine
Royal College of Physicians
6 St Andrews Place
Regents Park
London NW1 4LB

Employment Medical Advisory Service (EMAS): contact HSE 0541 545500 for local office

NHS Pensions Agency
Injury Benefits Manager
200–220 Broadway
Fleetwood
Lancs FY7 8LG

The National Aids helpline Tel: 0800 567 123

British Safety Council
National Safety Centre
70 Chancellors Road
London W6 9RS

Royal Society for the Prevention of Accidents (ROSPA)
Edgbaston Park
353 Bristol Road
Birmingham B5 7ST

Loss Prevention Council/Fire Protection Association
Melrose Avenue
Boreham Wood
Herts, WD6 2BJ

British Fire Protection Systems Association
4th Floor, Neville House
55 Eden Street
Kingston upon Thames
Surrey, KT1 1BW

Arson Prevention Bureau
51 Gresham Street
London, EC2V 7HQ

Association of British Insurers
51 Gresham Street
London, EC2V 7HQ

British Standards Institute (BSI)
389 Chiswick High Road
Chiswick
London W4 4AL

Trades Union Congress (TUC)
Congress House
Great Russell Street
London WC1B 3LS

ACAS: web site www.acas.org.uk, or phone nearest local office

Health & Safety Agency in Ireland (HAS)
10 Hogan Place
Dublin 2
Eire

Industry groups such as

Federation of Small Businesses
2 Catherine Place
Westminster
London SW1E 6HF

Forum of Private Business
Ruskin Chambers
Drury Lane
Knutsford
Cheshire WA16 6HA

Confederation of British Industry
Centre Point
103 New Oxford Street
London WC1A 1DA

Index